Adventures in Earth and Environmental Science

Teachers' Guide

Dr. Peter T. Scott

First released 2019 all rights reserved Felix Publishing

Felix Publishing 2019
www.felixpublishing.com.au email:
info@felixpublishing.com

Print copies available from publisher.

**ADVENTURES in EARTH and ENVIRONMENTAL SCIENCE
TEACHERS' GUIDE**

Also by the author:

ADVENTURES in EARTH and ENVIRONMENTAL SCIENCE Book 1

ADVENTURES in EARTH and ENVIRONMENTAL SCIENCE BOOK 1 PRACTICAL MANUAL

ADVENTURES in EARTH and ENVIRONMENTAL SCIENCE Book 2

ADVENTURES in EARTH and ENVIRONMENTAL SCIENCE Book 2 PRACTICAL MANUAL

ADVENTURES IN EARTH and ENVIRONMENTAL SCIENCE (the composite book containing books 1 and 2)

ADVENTURES in EARTH and ENVIRONMENTAL SCIENCE TEACHERS' GUIDE

ADVENTURES in EARTH SCIENCE

    A traditional Earth Science text incl. astronomy which
    is also available as a series of smaller books:

        Exploration Science (Field Geology and Mapping) Riches from the
        Earth (Minerals, Mining & Energy) Changing the Surface (Erosion
        and Landscapes) Rocks - Building the Earth
        Fossils - Life in the Rocks
        A Dangerous Planet (Earth Hazards)
        Through Sea and Sky (Oceanography and Meteorology) Beyond Planet Earth
        (Astronomy)

2019 Digital Edition   ISBN: 978-1-925662-07-8

2019 Print Edition   ISBN: 978-1-925662-06-1

Author: Dr. Peter T. Scott

All illustrations, photographs and videos by the author unless stated Cover photo: Andrew Scott of AJS Creative

Registration:

Thorpe-Bowker +61 3 8517 8342

email: bowkerlink@thorpe.com.au

No part of this publication may be reproduced, stored in a retrieval system, or transmitted in any form or by any means, electronic, mechanical, photocopying, recording or otherwise, without the prior written permission of the publisher.

© All rights reserved Felix Publishing 2019

About the Author

Dr. Peter T. Scott at his desk, 2008

Dr. Peter Scott is an award-winning Earth Science teacher of over forty years' experience Secondary schools. He holds a bachelor's degree, two Master's degrees and a Doctorate in both the fields of Science and Education. He has been an advisor in Science Education to several State governments. He has also lectured at UNE in Science teaching. Apart from his own research exploration, he has also travelled extensively and has visited many places of interest including North and South America, Antarctica, North Africa, the volcanic islands of the Pacific, Asia, northern Europe, Antarctica and Australia.

Table of Contents

BOOK 1.

Introduction	1.
Chapter 1: Exploration Science	18.
Chapter 2: The Blue Planet	32.
Chapter 3: Exploring the Rocky Planet	43.
Chapter 4: Rock-forming Minerals	62.
Chapter 5: Igneous Rocks - The Beginning	77.
Chapter 6: Sedimentary Rocks	86.
Chapter 7: Metamorphic Rocks	95.
Chapter 8: Weathering and Erosion	100.
Chapter 9: Environments of Weathering and Erosion	113.
Chapter 10: The Hydrosphere - Waters of the Earth	125.
Chapter 11: The Atmosphere - The Air Above	138.
Chapter 12: The Biosphere - Life on Earth	147.
Chapter 13: Energy and the Earth	163.
Chapter 14: Energy and the Sea and Sky	170.

BOOK 2

Chapter 15: Use of Resources and Energy	179.
Chapter 16: Economic Minerals	184.
Chapter 17: Non-renewable Fuels and Energy	194.
Chapter 18: Exploration for Resources	203.
Chapter 19: Mining Economic Minerals	217.
Chapter 20: Processing the Mined Ore	230.
Chapter 21: Monitoring and Management	238.
Chapter 22: Renewable Resources	254.
Chapter 23: Renewable Energies	272.

Chapter 24: The Earth in Motion	285.
Chapter 25: Volcanoes	293.
Chapter 26: Earthquakes	303.
Chapter 27: Wind, Rain and Fire	315.
Chapter 28: A Changing Climate	322.
Appendix A: Risk Assessment of Practical Work and Excursions	329.
Appendix B: Excursion Permission Note	332.
Errata from the First Edition	334.

Introduction

1. Why "Adventures" in Earth and Environmental Science?

The study of the Earth and its many environments IS an adventure! Studying Earth Science involves:

- **Exploration** - whether it be in remote areas or near places of habitation, field work usually involves the breaking of new ground (forgive the pun!). Often in the more remote locations, field work might be over ground which has never been as thoroughly explored as your study;

- **Exciting Places** - usually outside of the person's usual habitation range and often in very remote places in different parts of the world, in different physical and biological environments such as deep underground (mines and caves), on high mountain ranges and in different climatic conditions such as open oceans, glaciers and ice fields, hot and cold deserts and jungles;

A research station, Paradise Bay, Antarctica - mid summer

- **Meeting Interesting People** - who are often happy to share their own experiences and culture, not to mention their home, food and local knowledge. One meets a surprising number of people, mostly good people, such as fellow Earth and Environmental Scientists and students, field workers in the Earth and environmental Industries, Government officials; local land-owners and indigenous peoples;

- **Exciting Research** - especially when working on a new project or studying a new location, Earth and environmental research, as with all scientific studies can be especially motivating (sometimes to the point of happy obsession!). Just like a good detective story, there are things to find out, evidence to gather, step-by-step deductions to be made and finally a conclusion which answers the research question. Looking for gold or gems would be a simple example but it is when the study has a great number of research questions that the field and its companion laboratory work becomes exciting;

- **Studying at many levels** – whether it is a simple prospecting trip or a detailed study of the local environment, Earth and environmental science activities can occur at any age and at every level from the school student amateur to the professional scientist.

Whatever the nature of the study or the level of the student, the Earth is a dynamic and active place with exciting things to see and do. There is much to explore and many problems to solve before the Earth is a better place to live in.

2. Studying Earth and Environmental Science

Earth and environmental science covers a wide range of interesting topics, from volcanoes to exploring deep into jungles. The amount of mathematics, often common to Science subjects, has been reduced within the textbooks; providing adequate analysis of large amounts of data, but there is still much content to cover in order that a true understanding of the Earth and beyond is obtained.

Because of the amount of content needed to be learned, it is highly recommended that students adopt sound attitudes and practices. These include:

- **making good, concise summaries of each topic.** Whilst electronic devices such as computers and tablets are excellent for note taking, there is a tendency in modern education for students (and teachers) to upload or cut/paste previously prepared notes. These are good for gathering information as a primary source, but eventually a certain (minimal) form of summary will be required for study purposes and committing a basic set of notes to memory. After all, one must have some information stored in the brain (as well as on Hard Drive) to be able to use when using what has been learned. The best way is the ancient method of **reading, analysing and extracting the main ideas** and then **writing them down on paper** as a study summary. Doing this on a computer screen has some usefulness but it is not as good for learning as the hand-eye coordination which occurs with physical writing on note paper as a study sheet. Students should organize these sheets so that there is one page per major topic. The use of simple diagrams, charts and lists is an advantage to learning. For long sequences which should be known at all times and not simply retrieved from a data bank on demand, mnemonics are most useful. These may take a simple form of using the first letter of each word in the sequence to make another, simpler word which forms part of a crazy sentence. The crazier, the better. This is then learnt off by rote (try writing it out 100 times!) with the appropriate mental connections made to the real words and sequence. For example, consider the geological time scale of the Earth - Cambrian, Ordovician, Silurian, Devonian, Carboniferous, Permian, Triassic, Jurassic, Cretaceous, Tertiary, Quaternary:
Cold oysters seldom develop coloured pearls, their juices congeal too quickly.

American teachers may have to modify this as they recognise two major subdivisions in the Carboniferous – the Mississippian and the Pennsylvanian. Their mnemonic would then be:

Cold oysters seldom develop many precious pearls, their juices congeal too quickly

- **good study habits** and time management are not natural processes for most students, even those of mature age. A time-management grid or fixed calendar of regular study and assignment preparation could be suggested by the teacher. This would also include the necessary breaks for leisure time and social activities. Such a timetable should be flexible and include such well-known concepts that there should be social/leisure time immediately after classes. Study is best done just before sleeping. That a variety of different topics are best done in small amounts rather than a massive amount on the same topic, and that short (written) summaries of the nightly study are best done last. The latter concept does waste a lot of paper but the hand-eye coordination is most useful for memory retention.

- **completing practical work and assignments on time.** Never leave due work until the last minute. Students should be taught to plan a sequence of preparation e.g. gather notes by library/Internet research and reading, summarizing these notes in a cohesive form, planning the structure of the main work, then writing the assignment in one or several stages so that it is completed well before the due date. Assignments and practical work which occur regularly and often should be started as soon as possible after they have been set. Interruptions due to personal life will occur and if these interfere with the submission of work, extra time should be given by the teacher if they are valid.

- **noting down ideas**, words etc. not understood should become a regular habit, especially as data recording whilst working is part of the Earth Science method. Difficult questions, new observations and unknown concepts are the source of further research by asking questions. Teachers should encourage their students to always ask questions about what they see and do. There should be a classroom understanding that there is no such thing as an awkward nor silly question. We learn by questioning – even established ideas.

- **students should work independently and be accountable for their own work.** Although they may have to work and gather data in small groups, students must understand that the final outcome of this work should be their own. It is very easy using electronic media to copy and manipulate other people's original work. Even without computer programs to analyse student work for such plagiarism, it is usually obvious to the trained teacher to see that the level of understanding, terms used and data outlines is not typical of the age nor academic level of the student. Originality, no matter how simple, is the key to a good assignment.

3. How this Book Can Help?

The content of the textbook is up to date but based on over fifty years of interest in and study of the Earth and its many environments. It contains many photographs and videos, most of which have been taken by the author (unless stated otherwise) on location in many interesting places around the world - from the cold of Antarctica to the hot temperatures of the Nubian Desert; from the heights of the Andes and European Alps to the wide jungles of the Amazon.

The videos are not only used to show some of the most interesting places visited by the author, but also to show skills which cannot really be explained adequately in words. Students (and teachers) are encouraged to watch these videos and then research their locations as many videos show places which also have cultural and historical interest.

This Teachers' Guide also supplements the textbooks by giving practical advice based on over forty years of classroom teaching in both secondary schools and universities. It also provides curriculum and teaching suggestions including suggested practical work and demonstrations which can be used by teachers as a major resource in their classroom teaching.

4. A Need for Direction

Government Education Departments, District Boards and individual schools will have their own Earth Science syllabus and methods of directing staff in general terms how to prepare and teach the subject matter of each discipline. These guide lines are to be followed by the teacher in order to produce meaningful lesson plans.

In planning individual lessons, a teacher would probably follow a methodology similar to that on the next page.

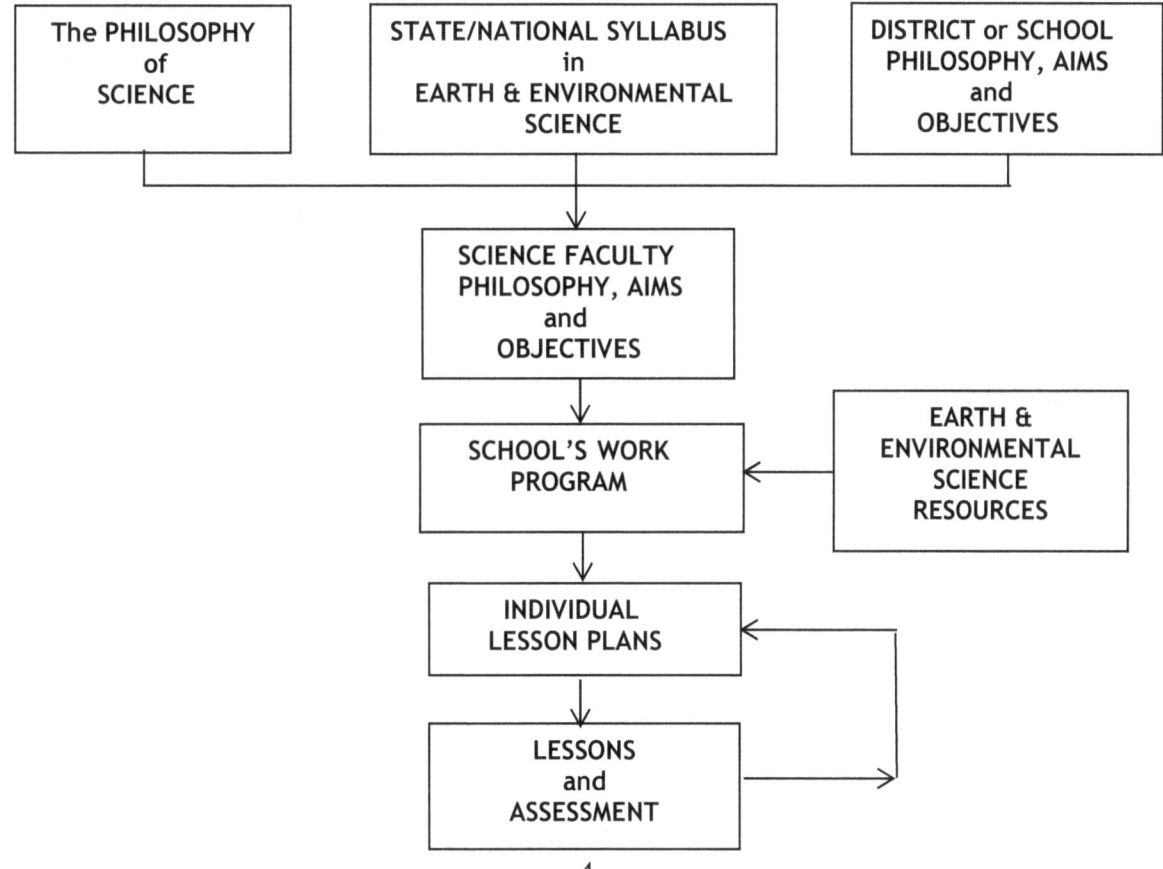

The important phase in this sequence is the structure of the teacher's own lesson plans unless he/she has the role of Earth Science Coordinator and must write the school's Earth Science Program. This will consist of a general preamble regarding the directions given by the State Syllabus and the school's philosophies and broad statements about the learning experiences and assessment requirements. The rest of the program will consist of a logical sequence of individual topic units which will probably each contain:

- the AIMS, THEME or other broad statement of the purpose of the unit;

- SPECIFIC BEHAVIOURAL OBJECTIVES which detail exactly what is expected of the student in a specific TIME (e.g. the end of the unit);

- The LEARNING STRATEGIES including practical experiments, excursions and demonstrations which are to be used;

- Any REFERENCES to the text book being used and other resources such as other books, web sites, computer programs etc. and

- the ASSESSMENT items to be used and their WEIGHT in the overall assessment.

It is important to include all of the material which is required from the National, State or District Syllabus without making the unit too long in length. The sequence is also important as both within the unit and over the entire program, the purpose is to develop a concept of the nature and methods of the science of the Earth in a progressive manner, with one section leading to the next so that units build upon prior knowledge and skills.

The program which is then developed should be a clear and concise document of intent which can be used for accreditation within school/state guidelines and also as a written statement for perusal by other teachers, faculty heads, principals and parents.

The student could also be advised about the necessary safety aspects of laboratory work and during field excursions. It is also useful to provide some assistance with study techniques and the writing of assignments and practical reports.

An example of a specific unit within an Earth Science program could be similar to the one shown on the next page:

Unit 1 - INTRODUCTION TO EARTH & ENVIRONMENTAL SCIENCE

SENIOR HIGH - YEAR11, Term 1   **8 LESSONS**

THEME: Earth and Environmental Science is a major discipline and a combination of several sciences. It follows the typical methods of scientific investigation common to all. Earth and Environmental Science relates to all of the systems of the Earth, the atmosphere, the geosphere, the hydrosphere and the biosphere. The Earth is dynamic; its systems often involve short and long-time frames and often violent processes involving the recycling of materials and energy conversion. Many environmental materials are useful resources and an economic use of these involves the need for a close examination of the total environment and a balanced viewpoint.

LIBRARY RESEARCH: Biography of one famous Earth and Environmental Scientists. Details are given at the end of this unit

OBJECTIVES: by the end of this unit, students should be expected to:

Knowledge
- K1.1 **Define** the sciences involved, namely geology, meteorology, biology, ecology, oceanography etc;
- K1.2 **Define** the terms 'environmental' and 'ecology' as it applies to the study of the Earth;
- K1.3 **Explain** the importance of the application of environmental principles to the study of the Earth's materials and energy with reference to the link between environmental sustainability and resource development;
- K1.4. **State** the purposes of each of the main study sub-divisions of Earth and Environmental Science;
- K1.5 **Identify** appropriate equipment and its use;
- K1.6. **Outline** the correct sequence and purposes of the scientific methods from:
    a) aim to conclusion and from
    b) hypothesis to law;
- K1.7 **Recall** some of the contributions made by famous Earth and Environmental scientists;
- K1.8 **List** a variety of Earth systems processes which are
    a) rapid and catastrophic and
    b) slow and continuous.

Working Scientifically
- WS1.8. **Develop procedures** for:
    a) observation with perception in both the laboratory and the field;
    b) performing experiments and
    c) recording observations (e.g. lists, tables, images, field relationships);
- WS1.19. **Deduce conclusions** about specific tasks and from several sets of data to show aspects of the whole Earth and Its systems.

Using Information Scientifically
- WS1.20. **Make simple calculations** from data obtained in introductory experiments and observations of Earth systems;
- WS1.21. **Infer** from numerical data and photographs, the dynamic nature of the Earth and its systems; (e.g. rates of erosion; volcanic events, earthquake data; populations, climatic conditions etc.);
- WS1.22 **Write and discuss reports** (practical/research) on the dynamics of the Earth's systems, environmental issues and the work of Earth and Environmental scientists using <u>appropriate styles</u> and within set time limits.

Skills
- S1.23 **Handle specimens and equipment** carefully and safely
- S1.24 **Use a hand lens** to examine fine detail of crystals and other specimen;
- S1.25. **Set up and use a binocular microscope** to look at fine detail of fossil plants and crystals;
- S1.26 **Sketch specimen or photograph observations accurately** to scale and with the use of colour and 3D perspective;
- S1.27 **Take simple measurements** such as scale on Earth photos, volume and mass measurements, and others in the laboratory and in the field

LEARNING STRATEGIES may include:
1. Examination of crystals, fossils and biological specimen with hand lens and binocular microscope;
2. Observation of some rock types to identify minerals, different environments with a variety of habitats etc.;
3. Introductory exercises on the Internet to locate data about the Earth and its systems;
4. Library research activity
   Some suggestions are: NICHOLAS STENO; JAMES HUTTON; GEORGES CUVIER; WILLIAM SMITH; CHARLES LYELL; ALFRED WEGENER; INGE LEHMANN; HARRY HESS; DAVID ATTENBOROUGH; JAMES LOVELOCK; ALFRED WALLACE; CHARLES DARWIN or others as appropriate.

REFERENCES
Adventures in Earth and Environmental Science Chapter 1.
General internet sites on Earth and environmental sciences (e.g. ASGIO, USGS, NASA, NOAA, BOM)
Career Documents and websites on Earth and environmental sciences

ASSESSMENT

Library Research (Biography)
Practical Reports
Part of Semester Test

Aims and Objectives - these two terms are often misused by people who do not have to use them as a practical guide within the classroom. Government educational authorities and School Boards are notorious for using organizational terms, policy statements and other long-winded expressions of what should be simple and helpful directions. Expressed simply, aims and objectives and similar directional statements are the outcomes which are expected of the student for a course, unit of work or lesson by a certain time:

> **Aims** are general outcomes, often expressed at the beginning of a course or unit of work. They may also be collectively expressed as a Theme or Purpose, Mission or similar set of general statements;

> **Objectives** are more specific and express the outcome required in terms of action verbs (e.g. describe, list, state, draw etc.) and a specific time by which these actions have to be achieved. These are the most useful descriptor of what teachers should do e.g.

> *"By the end of this Unit, students should be able to DEFINE the term Earth And Environmental Science"*

Objectives may be used as General Objectives with a broad definition and longer time range, such as for a complete unit of work or chapter of a book, or they may be Specific Objectives such as those detailed for an individual lesson. Students should always be made aware of the specific objectives, standards or task statements which they are to meet within a given time.

Aims and objectives should be in in simple terms and represent only one outcome unless this has several natural parts or have a number of specific criteria which sets a standard for this outcome. Objectives often represent outcomes to be tested by some form of assessment at some specific time, so they should represent only one, simple outcome to be achieved by a specific time. Vague statements such as to '*Know about Earth and Environmental Science*' is educationally useless.

5. More on Educational Objectives

Educational objectives often become an obsession with some academic educators. They are simply and literally a means to an end. Considerable time is spent by some in defining and classifying educational objects. For practical classroom use they should be clear in their purpose and concise. The use of certain key words is very practical when it comes to preparing individual lessons, observing students within the classroom and also for setting assessment items. For example, if a student can '*LIST the main sub-sciences within Earth and Environmental Science*', they can do so verbally in class and in an exam question which simply asks them to write these sciences down.

The author of this book has had a long experience in setting, classifying and using educational objects as a university academic, State Head of Syllabus in Earth Science, a field geologist, honorary Park Ranger, practical environmentalist and as a practicing classroom teacher for over forty years. What is suggested in this book works! Naturally, schools and their teachers are obligated to follow the guidelines for writing objectives provided in any syllabus or other official guidelines and putting these ideas into practice within the classroom, but teachers have a wide range of professional choices and relationships with which to implement these guidelines. The Practical Manuals which supplement the textbooks on EARTH and ENVIRONMENTAL SCIENCE are good example. There may be some mandatory practical activities prescribed by the higher educational authorities and these have been included. Other practical activities can be done in total and in the sequence given, or more likely selectively chosen or modified to meet the student's abilities, resources and needs. Flexibility in using these resources is essential in good classroom teaching.

Many classification systems of education objectives often derive from that of Benjamin Bloom and others (Bloom, B.S. Edit., Engelhart, M.D., Furst, E.J., Hill, W.H., Krathwohl, D.R. (1956). *Taxonomy of Educational Objectives, Handbook I: The Cognitive Domain*. New York: David McKay Co Inc.), who devised a taxonomy of objectives in the Cognitive or Knowledge Domain. This was later extended for the Psychomotor (or skills) Domain and the Affective (or values and attitudes) Domain. Each domain can be further subdivided into appropriate categories depending upon needs. There are many such systems and the one given in the example (above) has a simple classification into Knowledge (Cognitive Domain) Working Scientifically, Using Data Scientifically and Skills (all in the Psychomotor Domain) but no listing of objectives in the Affective Domain. This was done for practical reasons as writing objectives concerning attitudes, interests and values can be awkward to write and difficult to honestly assess.

There are some key terms which are useful in devising educational objectives. Using the taxonomy used in the example (above) such terms could be:

Knowledge Objectives	Key Terms for Learning Strategies and Assessment
To recall knowledge	List, define, state, describe, select, identify, recognise, categorise, illustrate, tabulate recall facts, terms, symbols, formulae, procedures, theories reproduce shapes, patterns and diagrams

To understand	Perceive, describe, classify, explain, discuss, summarise, interpret, interrelate, compare, contrast, evaluate, exemplify, substitute, reconstruct:
	    ideas, themes, issues, laws and principles
	    symbols, patterns, pictures
	    formulae, numerical magnitudes

To apply knowledge and understanding by	Deduce a result, solve a problem, draw a conclusion calculate a value

using previously learned concepts to	infer a relationship predict an outcome propose an hypothesis, eliminate an incorrect hypothesis propose a sequence of events, explain a pattern or feature estimate a magnitude or approximate solution

Working Scientifically (Skill) Objectives	Key Terms for Learning Strategies and Assessment
To design and implement investigations	Recognise the problem and relevant issues propose hypotheses or tests for hypotheses plan an investigation, design experimental procedures identify and control variables select appropriate techniques for laboratory and field studies identify safety issues and procedures record observations using a variety of methods select appropriate measuring devices, scales and units identify errors in measurement describe properties and changes classify objects using appropriate properties distinguish relevant from irrelevant information distinguish observations from inferences

Using Information Scientifically (Skills)Objectives	Key Terms for Learning Strategies and Assessment
To use information scientifically to:	Interpret identify trends or anomalies interpolate and extrapolate rank analyse, infer and predict relate cause and effect describe relationships (qualitatively and/or quantitatively) compare and contrast propose a conceptual model present information in a variety of modes and forms use secondary data sources use appropriate conventions of language and referencing use suitable genres

Psychomotor Skill Objectives	Key Terms for Learning Strategies and Assessment

To use appropriate equipment and procedures	Use items of standard laboratory equipment safely – stock solutions & chemicals, heating apparatus, glassware Use optical equipment with care – microscopes, telescopes, spectroscopes, hand lenses Use field equipment with safety – geological hammer, compasses, maps, GPS units, cell phones & apps. Navigate a given territory with safety Carryout field work regarding climatic, surface, biological and other hazards Use computers, tablets and cell phones appropriately Apply basic First Aid and Emergency Procedures correctly when appropriate

Given time and some thought, Affective Domain objectives can be written but they are often of a vague form such as "SHOW INTEREST in the study of minerals." Or "HAVE A POSITIVE ATTITUDE towards the use of natural resources". Exam questions on such objects can also be difficult to write and harder to assess accurately. Determining whether or not a student shows interest, or has the right attitude to a specific topic or activity is best left to teacher observation. Certainly, some teachers have used open-ended questions, essays on controversial Earth topics or even 1-5 rating scales to obtain some value of a student's mental attitude, however these can also measure the student's bias due to home or media or what is perceived to be trendy or politically correct. Leaving the Affective Domain out of school-based assessment seems to work well. However, if the Affective Domain is to be part of the objectives/assessment system, then some key terms could be:

Affective Objectives	Learning outcomes
To develop positive attitudes and values in or towards	personal behaviour in terms of cooperation in groups tolerance of individual differences, religious and cultural diversity being open-minded and critical of data, being skeptical and willing to shift in the face of evidence, being systematic and persistent in pursuit of understanding. Suspending judgments until all data has been gathered, admitting the strengths and limitations of science having a balanced view as to the use/abuse and protection of the environment

6. More on Lesson Preparation for the Novice

As a Head of Department within large schools, as a member of a State Review Panel which looked at school items for accreditation and as a Science Methods Lecture to Trainee Teachers, the author has often found that many classes were often taught by teachers with minimal and sometimes no Earth and Environmental Science training at all. This usually was due to the lack of properly-trained teachers in the Earth and Environmental Sciences. More often, the teacher may be well-trained in one or several sciences such as Geology and Chemistry but has had limited training in others such as Biology, Meteorology and Environmental Science. It is very common for good Biology teachers to be given an Earth and Environmental Science class with short comings in the massive amount of Geology, Meteorology and Oceanography required. Good teachers then go about learning the correct details of the deficient knowledge and skills. Others fall back on their old high school ideas. Others will make full use of textbooks and associated resources such as those of the EARTH and ENVIRONMENTAL SCIENCE package.

It is outside of the scope of this book to give a detailed program of teacher training but in case the reader would like another opinion on lesson planning, this is briefly given here. The system was derived from the military and used in the training of their instructors. It was modified by the author (formally Commissioned into the Infantry) and used in establishing Science Teaching at a well-known country university.

Lesson planning is the final stage of the teacher's translation of the topic into a logical sequence of meaningful understanding for the students. At the beginning of a teacher's career or for special occasions, considerable time may be needed in preparing specific lessons. These first steps and the resources obtained for them should be kept in the teachers file for later use. As the teacher becomes more experienced and has developed a teaching strategy for each lesson within a well-rehearsed program, the preparation time becomes less but never unimportant. For special occasions, such as School Open Days, Science Fairs and teacher performance inspections, detailed lesson plans are most useful.

Every lesson is different because of the dynamics of the classroom or lecture hall, so it is important that the teacher has prepared notes, rehearsed audio-visual presentations and practical activities which have been prepared, tested and evaluated for safety. A well-organized Teachers' Diary with an outline for the lesson is usually the main written resource of an experienced teacher.

Every lesson has three parts -

- the **INTRODUCTION** which includes a **connection** to the previous lesson (perhaps a few questions around the class or answers to a home assignment), a **motivation step** to gain attention (e.g. an interesting mineral specimen, a spectacular photo, a cartoon or a short video) and a rationale (or reason for doing well in this lesson);

- the **MIDDLE** or body of the lesson which contains a sequence of steps - a script possibly - of the main part of the lesson. It will contain all of the reminders that the teacher needs. For an experienced teacher, these will be brief but a new teacher may expand on them with additional notes on a facing page in the diary; and

- the **END** which will follow the final step of the last student activity with some assessment of the success of the lesson. This might be a few questions to selected students or short written notes or testing of a simple skill. There may be some follow-up in the form of completion of some unfinished task of the lesson at home or some simple research required for the next lesson.

One important aspect of teaching Earth Science is the need to have a good collection of resources at hand. If the teacher has the luxury of a personal classroom, then a good collection of specimens, equipment, charts, maps and models could be put on display and ready to hand. Also, an electronic collection (Hard Drive or USB stick) of suitable photos, videos, worksheets and additional notes is useful for downloading to students. The textbooks, ADVENTURES in EARTH and ENVIRONMENTAL SCIENCE, Parts 1 and 2, have a great many linked videos, mostly made personally by the author in the field, which can be copied in to the URL bar of any tablet or computer. The electronic versions (Kindle format using the free Kindle App on any tablet or computer, or using a host server for the eBook) will have direct, clickable links to these videos.

An example of a simple lesson plan is given on the next page. Note that it is a concise script rather than a lengthy set of notes. This page might be on the left-hand-side of the teacher's folder or diary and any additional notes would be on the right-hand-side. Alternatively, notes, photos and videos may be linked directly in the teacher's electronic notebook. These resources may also be linked to student electronic notebooks for their download. Items in red are comments on the sections of the plan and not part of the lesson plan itself. A sample lesson plan is given on the next page:

BEGINNING of the lesson plan (may have a reminder of last lesson after TOPIC)

WEEK: 1. **DAY:** Monday **DATE:** Feb.4 **TIME/CLASS PERIOD:** 4

CLASS: 11 EES 2 **PROGRAM UNIT:** 11.1
TOPIC: Introduction to Earth & Environmental Science

MOTIVATION: Items of equipment on desk e.g. large crystals, geology pick & compass, barometer, binocular microscope & salt, sextant, water testing kit (if available).

OBJECTIVES K1.1, K1.4, WS1.8 a & c, S1.23, 1.24

RATIONALE: A study of Earth & Environmental Science is useful in knowing and protecting our environment and using its resources wisely.

PREPARATION/RESOURCES: Motivation items, hand lenses, sheets of paper, class set of large quartz aggregate crystals.

MIDDLE part of the lesson plan

LESSON SEQUENCE:

1. Welcome, seating arrangements/group organisation, settling of class;
2. Reference to Junior High Science program (perhaps). *Connecting link from past lessons*
3. DISPLAY and intro. to the items on the front desk.
   What are they used for?   Who would use them?   Why?
4. EXPLANATION from students' answers
5. DEMONSTRATION use of each item generally. Salt crystals under binocular microscope. Ham up poor use of hand lens - students to correct teacher.
6. STUDENT ACTIVITIES:
    1. Hand out class set of crystals and paper (pencils if students not equipped)
    2. Students asked to sketch in 3D the crystals. Names of student on sketch.
    3. When finished, individuals may go and view through the microscope and carefully handle other items (Watch!)
    4. Students return to seats and hand in sketches.
    5. Class judging of sketches (no names given) 1-10 rating (perhaps a reward of sweet for top students - with other smaller sweets for the rest for their efforts)
7. QUESTION: What sciences do these items come from? (broadly from their general knowledge and Junior High program)
8. LISTING on chalkboard or screen - of the main science headings (e.g. geology, oceanography, meteorology, ecology)
9. EXPLANATION: these sciences are linked together as EARTH & ENVIRONMENTAL SCIENCE - give broad definition. Earth systems and spheres.
10. STUDENT NOTES Intro and copy diagrams with main headings. Students to write own definitions of sub-sciences.
11. DISCUSSION of student definitions - any corrections made on board/screen

END of the lesson plan as an assessment of the lesson objectives and prelude to the next lesson

ASSESSMENT: 1. Questions to class e.g. What are the main sciences involved with Earth & Environmental Science? How are they defined? What are they used for? What are the main spheres of the Earth? Define each.
   2. Use as a later test item
NEXT LESSON: REAL SCIENCE! Training to sketch crystals scientifically and the microscope

Earth and Environmental Science, like Biology, Geology, Oceanography and Meteorology is a field-based discipline. That is, it uses the methodology of Science to explore a wide and diverse environment and solve problems relating to the world beyond the traditional laboratory. Laboratory work is very important also as the Earth and Environmental Scientist will spend considerable time within an institution preparing field studies and then analysing data from them or remotely-sensed instruments.

The teaching of Earth and Environmental Science should reflect these principles but it is difficult in both Secondary and Tertiary teaching spheres to do this because of the youth and inexperience of the students, the vast amount of theoretical knowledge to be imparted and the limitations of distance, safety and cost.

Some concept of the practical nature of Earth and Environmental Science can be achieved by integrating regular set practical activities and teacher demonstrated supported by good use of audio-visual material and the Internet.

The text book ADVENTURES in EARTH and ENVIRONMENTAL SCIENCE tries to make the study of the Earth by the use of many formal and informal practical activities in the classroom by way of the companion practical manuals which are matched to the textbooks. As psychomotor and communication skills are major objectives in this program, students should be able to use appropriate equipment, have good powers of observation, a well-developed sense of enquiry and the ability to express themselves both formally and informally in speech and in written text. Formal practical activities and the reports generated from them are designed to develop those skills which will be needed in later years by researchers and field scientists.

The companion books ADVENTURES in EARTH and ENVIRONMENTAL SCIENCE – PRACTICAL MANUAL 1 and 2 contains formal activities which may be used as part of the school's teaching and assessment programs. The activities given in these manuals are matched to the chapters and their content of the textbook. Additional helpful notes and some supporting demonstration activities are given in this Teacher' Guide in the appropriate sections relating to the chapters of the textbook.

It would be hoped that the school would have sufficient resources so that the Earth and Environmental Science staff can use these activities regularly with good support from audio-visual and computer resources. Earth and Environmental Science is NOT a chalk-and-talk, teacher-dominated subject. Every lesson should be an adventure with something interesting to examine or discuss. It should involve ideas similar to the enquiry and thirst for knowledge of the early pioneers, the love and understanding of even the smallest detail of the land known by indigenous peoples and a wise understanding and laws of the sea and sky. All this is encompassed in the philosophy and methods of modern Science.

Students who are using electronic notebooks in the classroom will have to complete any diagrams or sketches by hand on blank paper in such a clear manner (all students might enhance their sketches by going over the outline and main detail in heavy pencil or black ink) that they can then be photographed on site or scanned electronically at home for inclusion in their notes or practical reports (which may/may not be submitted electronically, depending upon the teacher's policy).

7. Safety in the Classroom Laboratory

Experience has shown that some Secondary and Tertiary students are uncoordinated and socially irresponsible. For this reason, they and rest of the group need to be protected by the application of a set of firm but fair Safety Rules. These should be simple, able to be comprehended by the student and practiced. It is also a good idea to have them posted within the classroom. In time, most groups get into a working pattern and so The Rules become simply standard learning procedures and so a good, working social atmosphere develops. An example of such rules is given below:

SAFETY in the LABORATORY

The following policy is in place to ensure that the laboratories are safe places for students to work in. Respect for self, others and property is first priority. Please ensure that:

1. Safety apparel is to be worn at all times - aprons, good enclosed shoes, eye glasses etc.;
2. you do not enter a laboratory without teacher supervision;
3. unauthorized experiments are strictly forbidden. Variations in procedure must be approved;
4. food and drink are not consumed in the laboratory;
5. movement and noise should be kept to a minimum as distractions cause accidents;
6. items and liquids are not to be thrown at any time;
7. you do not touch, taste or smell chemicals and specimens unless specifically directed by the teacher;
8. spillages, breakages and accidents are reported immediately;
9. you know where all safety switches and equipment are located and how they are used; and that
10. the work area is clean and tidy. When finished, clean apparatus and return with chemicals etc. to appropriate place. Wipe down laboratory bench and wash hands. Soils and biological specimens must be disposed of into containers allocated for their disposal.

In addition, use of electronic equipment such as computers, tablets and cell phones should be appropriate for the learning environment and secondary to the teaching process. Some schools ban the use of cell phones but with a few simple courtesies they can be useful in the classroom.

8. Safety in the Field

This requires more vigilance by the teacher than in the protected laboratory environment. In the field, away from the classroom, there are other external influences such as:

- the climate, which could be excessive in its heat or cold, windy, wet or subject to sudden change;

- the terrain, which may be steep, rugged, slippery, loose, and full of ravines, caves or mine shafts;

- the vegetation, which may be dense, tangled, thorny and sometimes poisonous; and

- the wildlife, may be dangerous if disturbed or generally a nuisance (such as flies or mosquitos).

The students themselves sometimes may be a major safety problem if they:

- wander about and not keep with the group. There is real danger of their becoming lost, falling into a ravine, mine shaft, waterway or other dangers;

- are unprepared for the trip, especially if they have failed to be advised or have ignored advice about such necessities as:

    CLOTHING - especially adequate footwear, head covering, suitable clothing for the climate and sturdy boots;
    WATER and FOOD as appropriate for the trip;
    SPECIAL SAFETY GEAR (e.g. walking poles);
    INSECT REPELLANT and SUN PROTECTION;
    STUDY ITEMS such as Excursion Guide, notepad and pencil.

- fail to forewarn about special needs such as allergies and other problems involving outdoor activities; and

- are likely to indulge in foolish behaviour such as throwing stones or being a distraction.

Sometimes the teacher may be at fault due to a lack of thorough preparation and knowledge of the area (which should have been recently scouted). A light-weight First Aid pack, some spare writing pads, pencils and handout notes should also be taken. Dangerous obstacles should be avoided and only safe, secure tracks should be used. Good navigation is essential and the proposed route and time of return should be left with any local authority such as local police or park rangers as well as with the school authorities.

Early teaching of good field habits should lead to more extensive training for the potential Earth Scientists to cope with extremes of climate and terrain. Living and traversing difficult country, often alone or in a small isolated group, is sometimes the main working method of the field scientist.

A typical set of Field Safety Rules which is given out to the students below legal age is given below:

FIELD RULES

When in the field:

1. Listen to all instructions - especially about specific local hazards;
2. Keep with the group - do not wander;
3. Do not enter bodies of water unless told to do so;
4. Do not enter old mines, industrial workings or caves unless told to do so, and then with caution;
5. Do not climb cliffs nor stand under or near unstable rocks;
6. Do not throw any objects, especially hammers or rocks;
7. Watch your step, especially on slopes and in close vegetation;
8. Wear appropriate field clothing at all times. Be prepared for sudden changes in the weather (rain, cold/heat). Carry a waterproof jacket;
9. Watch out for traffic when on or near roads and railway cuttings;
10. Carry own water and some food;
11. Keep movement and noise to a minimum and leave the environment as it is found;
12. Do not use cell phones inappropriately. Headphones are not allowed. Take own care of cameras.

I have read and understood the above Safety Policy and agree to abide by it for the safety of myself and others.

_____ _____ _____
Print Student Name   Class/Group   Date

_____   _____
Students Signature   Read and witnessed by
                                            parent/guardian

Whilst such a document may not be considered legally binding in many states, it does give the student and their guardians some idea of the standards of behaviour which are considered to be minimal. If the institution has a policy against the use of cell phones in the classroom, this is one instance where they could be useful, however many field areas which are visited may not have good reception. For this reason, the organiser of the field trip should look to a secondary source of safety communication. Certainly, the activities, locations and times of the field trip should be well-known to any local authority, such as Rangers or police units with a given time after which assistance will be needed. Well prior to the event, the field trip should have had approval from a higher authority within the institution. More will be given on safety and useful hints in the field in reference to each chapter of the textbook.

Chapter 1: Exploration Sciences

1.1 Theme

This chapter explain how Science works using Geology as the example. The branches of Earth and Environmental Science, the Scientific Method, some basic skills and some of the simpler statistical and mathematical procedures needed to analyse data honestly are explained.

1.2 Rationale

This is the introductory chapter to the Earth and Environmental Science and of the philosophy, methods and the basic mathematics used within Science generally. Students should develop the habits of the Scientific Method such as accuracy of observing, measuring and recording data; evaluating and measuring the accuracy of these measurements; and an open honesty in coming to logical conclusions. They should also continue developing their curiosity and interest in how the Earth and its systems work.

At the very beginning of this chapter, students should be introduced to the usefulness of Science as a whole; both as a possible career pathway, but also as a personal interest and (often lucrative) hobby. Careers can be undertaken at the university level as scientists in any of the many branches or in the many trades and ancillary professions which are also vital to Earth and Environmental Science activities. Some careers advice from literature of visiting experts would be of use at this time.

1.3 Notes on the Practical Work (see PRACTICAL MANUALS 1 & 2)

Student practical activities are given in the companion book ADVENTURES in EARTH and ENVIRONMENTAL SCIENCE – PRACTICAL MANUAL BOOKS 1 and 2. Suggestions and comments of these activities and any appropriate guidelines are given in this book within each topic chapter. Students would be expected to read and follow the procedures given in the Practical Manuals and then write up the practical work for submission as required. Teachers may prefer to use the practical activities at the beginning of the Earth Science program as examples and any assessment may/may not become part of the ongoing grade system. Mandatory practical activities from the syllabus are included but other practical activities should be used according to the planning of the school and teacher. Practical activities which are not mandatory may be excluded, modified or replaced with others. They have been written for a normal school situation with typical laboratory equipment and funding based on the author's own experiences in schools.

<u>Experiment:1.1</u>: Introduction to Systems

This first activity is designed to teach students about the concept of systems when looking at a broad area of study, especially the differences between an Open System in which both energy and matter can transfer and a Closed System in which matter is conserved. In this activity the AIM, METHOD etc. can be simply copied out by the student when writing their report as it is given here written in the third person, past tense and in point form. In later activities, the method and other instructions will be in normal text (future tense directions) and the student will be required to interpret the instructions and questions to and rewrite their own reports in the traditional style using third person, past tense and point form (or as required by the faculty) as well as their own natural sketching skills to reproduce a sketch of the apparatus. Students should also be shown how to use a balance to accurately measure mass and how to judge the instrument's error.

In addition, the teacher may also request that the students do some additional research on the Internet or from the text such examples of closed and open systems and how to distinguish between them. Not all experiments will have this research component and teachers should feel free to use it, perhaps as homework or part of the write-up report. Alternatively, the research component may be ignored for some experiments.

QUESTIONS from the Experiment:

PART A:

1. **Was there any matter lost from the system? If so, what was it?**
   Yes. Gas was lost from the system (Students should be reminded about the reaction of carbonate with the acetic acid of vinegar).

2. **Was there any heat lost or gained from the system? If so, what type of reaction was it?**
   Heat was given out as this is a slightly exothermic reaction. (If it is too small to be felt, a demonstration with marble chips and dilute hydrochloric acid should show the same effects).

PART B:

1. **Was there any matter lost from the system? If so, what was it?**
   No. The gas was retained in the balloon.

2. **Was there any heat lost or gained from the system? If so, what type of reaction was it?**
   As it involved the same reaction, heat was again given out as a slightly exothermic reaction.

<u>Experiment:1.2</u>: Introduction to Scientific Observation: Examination of a Fossil Specimen

This is an exercise in observation of detail and sketching. Students should be encouraged to look for detail, especially patterns or recurring shapes.

Students should be given some explanation in such skills as:

- **How to sketch** experimental equipment, specimen and some field relationships. Photographs are also important but sometimes conditions do not allow some of the finer detail. Moreover, sometimes it is important, especially with specimens, to have a representation showing the main features rather than a real-life reproduction which may only show a few.

When sketching a specimen, students should be encouraged to

1. suspend judgment and examine the whole set up or specimen carefully, using all of the senses possible supported by items such as hand lenses and binocular and geological microscopes;

2. look at the overall shape of any specimen – its edges, internal features, how any components (e.g. crystals or grains, cells or organelles) are fitted together and their relative orientation; colour and shading as well as any impurities or other variation.

3. draw an outline of any specimen. This may be a series of straight lines, regular solids (e.g. cubes, rectangles etc.), curves or rounded shapes. Draw this outline to a respectable size. This may be several times greater than the real specimen, especially if it is viewed through a microscope.

4. add any internal features seen. These may be lines of cleavage, cracks, bubbles, crystals, internal parts or anything other than a uniform surface which is also possible.

5. give any specimen a three-dimensional effect by adding some darker sides and edges. This is unlike the traditional 2D scientific drawing used in sketching apparatus.

6. use coloured pencils (less destructive than inks or paints!) to colour any specimen as accurately as possible – apparatus can usually be left as black and white line diagrams. With specimens, the colours can be smeared with a dry finger or piece of paper and then go over outlines and main features with heavier pencil and finally

7. compare the size of the sketch to the original specimen and estimate the enlargement (or reduction) factor as a SCALE (e.g. x 2 or x ½ etc.).

- **How to use a hand lens** is one of the basic skills of a field scientist. Simply, the lens is put near the preferred eye (but keep both eyes open to prevent strain) and the specimen is brought up to the lens.

- **How to use a Binocular Microscope.** These are also called Dissecting Microscopes and are often used to observe specimens in the Biology lab. A set (of say 10) is useful in the Earth Science laboratory or they may have to be borrowed from the Biology Department.

They have a flat base (often inter-changeable as white or black for contrast) and are illuminated from above by natural light or an appropriate lamp. Specimens can be examined by:

1. Adjusting the two (binocular) eyepiece tubes to suit ones eye positions.

2. Moving the main barrel UP to give the greatest space for the specimen.

3. Placing the specimen carefully onto the centre of the base.

4. WHILST LOOKING FROM THE SIDE, move the main barrel down until the end lenses (the objective lens) are ALMOST touching the surface of the specimen;

5. Whilst looking through the eyepieces, rotate the focusing knob so that the main barrel MOVES UPWARD AWAY FROM THE SPECIMEN (emphasize that focusing is done upwards) until a clear focus occurs. This may have a variable DEPTH of FIELD requiring different focusing of different parts of the specimen if it is very three-dimensional. Watch out for the objective lens so that it does not touch the specimen.

6. To estimate scale, measure the width of the field of view (i.e. the circle seen when viewed through the eyepieces) by placing a ruler under the microscope once that the specimen has been removed. Give the scale as a length (in millimetres) across the side of the sketch. A good idea in sketching what is seen is to draw a circle (to contain the sketch) which represents that which is seen through the eyepieces.

CONCLUSIONS from the Experiment:

1. **What are the main distinguishing features of the fossil?**
   This will depend upon the type of specimen, but hopefully students can describe the venation on any leaf or pinnule seen. Note that some books will refer to pyrite crystals as the specimen. These are also good to use with the detail being seen as cubic crystals and striations on their faces.

2. **Comment on factors which might limit the accuracy of observation?**
   Here the main errors involve the ability to use the microscope focus and the student's power of observation. Dirty lenses and poor specimens will also reduce good observation.

3. **Why is it important to have an accurate scale in drawing?**
   A sketch on paper will not show the true size of the specimen so a scale is needed. The best way to judge the scale is to place the specimen near the sketch and then estimate the fraction or multiple of size. For small specimens greatly magnified, their size can be measured against the field of view previously determined with a view of a scale on a ruler or grid.

Experiment:1.3: Measurement Exercise: Size of Feldspar Crystals

This is a paper exercise assuming that the school will not have sufficient specimens of basalts with different sized crystals. It also is the first exercise in which students will have to read, interpret and then write their own version of the experiment using the required format of third-person past-tense etc.

CONCLUSIONS:

1. **What was the average size of the crystals for each of the depths and their error of measurement?**

    This will depend upon the student's accuracy, especially with Photomicrograph 1. Which has very small lengths, perhaps they could use a hand lens? They should be able to measure about 20 lengths for each.

2. **What type of error of measurement is this?**
    This is an analogue measurement using a ruler scale so the instrument error will be one half of a millimetre.

3. **Is there a relationship between crystal size and depth of their formation? If so, what is it? Can you express this relationship as a mathematical expression?**
    The relationship is a direct one i.e. crystal size increases with increase in depth, due to a slower rate of cooling. Mathematically one can say that length (l) is proportional to (α) to depth (d) or:

    $$l \; \alpha \; d$$

4. **Are your measurements made on the photographs a true representation of the sizes of the crystals? Why? How can you improve on the method of randomly sampling which crystals are to be measured?**
    This is a matter of sampling. It should be a reasonable set of measurements for photomicrograph 1 with many crystals, but errors of measurement, but validity might be stretched with the other two. Additional sampling by moving the original slide about would show if the photograph was a good representation of the crystal sizes in the specimen.

5 & 6. **Other conclusions and testing the hypothesis: that size of crystals depends on depth.**
    Comments could be made about the uniformity of the crystal size in each photomicrograph and additional photomicrographs at other depths or other well-shaped crystals (e.g. hornblende) in other igneous rocks at different depths could be viewed to test the hypothesis.

Additional research: (Optional at the teacher's discretion)

What are feldspar crystals? How do they form in rocks such as basalts? Where would VERY large crystals of feldspar (say a metre long) form? What use is feldspar in daily life?

Feldspar is used in glasses, ceramics, as fillers, enamels and glazes, paints, in the plastics and rubber industries, as fluxes and as gemstones in the crystalline forms.

See http://www.ima-na.org/?page=what_is_feldspar

Many photomicrographs are available on the Internet but actual viewing of a glass slide thin-section through a petrological microscope fitted with a camera and shown on a large screen is ideal. If one has a point-counter attached this is even better. If a petrological microscope is beyond the expense of the school, a common biological microscope can be used with thin-sections by placing a piece of Polaroid plastic below the stage and a smaller piece within the eyepiece barrel.

Polarizing plastic sheets are an excellent accessory to have in the classroom for several purposes. These can be purchased at several supply companies, e.g.

https://www.amazon.com/Polarizing-Film-Sheet-Gadget-Electronics/dp/B004X3XFHU

Experiment:1.4: Using a Topographic Map

This is another worthwhile paper exercise. Again, students are encouraged to write paper or electronic reports but here there is a drawing skill component. Students who are not adept at electronic art should do it on paper. An alternative is to do the art on paper, highlight it in black pen, scan it and then insert it as an image in the electronic report.

Modern Earth and Environmental Scientists have a great range of computer mapping programs and in the field one can upload field data which has been recorded electronically via satellite link and have the map and cross-sections constructed at Head Office at a great distance away. However, as in many scientific fields, it is good practice to know how topographic maps can be used and how cross-sections across them can be drawn.

Electronic mapping and navigational devices are extremely useful but they can suffer in extreme conditions of field work, from lack of connectivity and from internal malfunction and loss of battery power. Printed maps, and accompanying aerial photos, are still very useful in the field and in the laboratory.

In this exercise, the map given in this book could be substituted for a class set of local topographical maps if in a rural area. Questions given in this book can then be reset for local conditions. The map given here is of a coastal area and the hills and ridges are mostly of sand.

QUESTIONS from the Experiment:

PART A: The Topographical Map.

1. **What is the Scale of this Map? What does this mean in reality?**
   Scale is 1:25 000.

This means that one centimetre, or inch, on the map represents 25 000 centimetres or 250 metres, or approx. 700 yards on the ground.

2. **What are the GEOGRAPHICAL COORDINATES (i.e. Latitude & Longitude of this area?**
Always give the latitude and then the longitude. Grid references and map names etc. are useless in emergency so use latitude & longitude. Here, they are 27 degrees (⁰) and 30 minutes (') South latitude and 153⁰ 25' East longitude.

3. **Use an atlas (printed or electronic) to find where this place is located.**
This map is a fictitious map but would be located offshore from Brisbane, Australia.

4. **What is the MAGNETIC VARIATION of Utopia Point? (as degrees East)**
Remember that Magnetic Variation is the angular difference between True (Geographic) North and Magnetic North. On this map (it varies from location to location), that difference is 0.5 + 10.5 or 11 degrees East. There is only 0.5 of a degree difference to the man-made Grid North but if one was on the ground using a compass to sight a bearing across which to walk, one would have to ADD this EASTERLY variation or if drawing a line of walking (a traverse) on the map and then determining a compass bearing along which to walk, one would have to SUBTRACT this variation. It gets complicated: if one has a WESTERY variation then all of the procedures given above are reversed!

5. **List the GRID NUMBERS which represent:**
    a. **the EASTINGS.** These come first and are those numbers on the horizontal axis going East (e.g. 39 to 42) **and**
    b. **the NORTHINGS.** These are the Northward bearings on the vertical axis (e.g. 57 to 60)

6. **Use the SCALE and a ruler to measure the length of the JETTY at UTOPIA POINT.**
From where it starts at the road, it is approximately 540 metres long.

7. **Use a piece of string to estimate the length of the coastline shown on the Map.**
In extreme circumstances in the field (Oh no! No string!) – use bootlaces. Placing the string along the coast and following its outline (not the sandbars!), it is approximately 6500 metres (6.5 km) but this will vary slightly depending upon where it is measured. Very useful when determining distances walked down a stream.

8. **Within which GRID SQUARE is HOSPITAL POINT?**
3957, always taken from the bottom left of the square.

9. **What objects are located at GRID REFERNCE:**
    a. **393583** Polka Point
    b. **428590** the centre of Blue Lake

10. **What is the GRID REFERNCE for:**
    a. the end of the JETTY at Utopia Point
        i. 394576

b. the lone hill (118 m high) in the far north-eastern part of the map
   i. 422597

11. What is the CONTOUR INTERVAL of this map?
    10 metres above sea level

12. What is the feature between grids 413576 and 414564?
    It is a long ridge which is probably a frontal dune parallel to the coast.

PART B: The Cross-section

Reference: Page 34, ADVENTURES in EARTH and ENVIRONMENTAL SCIENCE.

Having drawn a cross-section box to the appropriate scale (in this activity, the Vertical Scale = the Horizontal Scale i.e. Vertical Exaggeration = 1), and transferred the data from the line A-B on the map, vertical lines are taken (as working which can be erased later) down to each of the appropriate heights above sea level. A hand-drawn line is then used to connect these intersections (land surfaces are NOT like economics graphs!) to obtain an approximation of the land surface. Note that assumptions are made at the start (where the traverse AB is below 700 metre height and also near B where it will be below 1500 but above 1400 metres above sea level. In these places, dotted lines are used to show the level of uncertainty.

Students' cross-sections should look something like that shown on the next page:

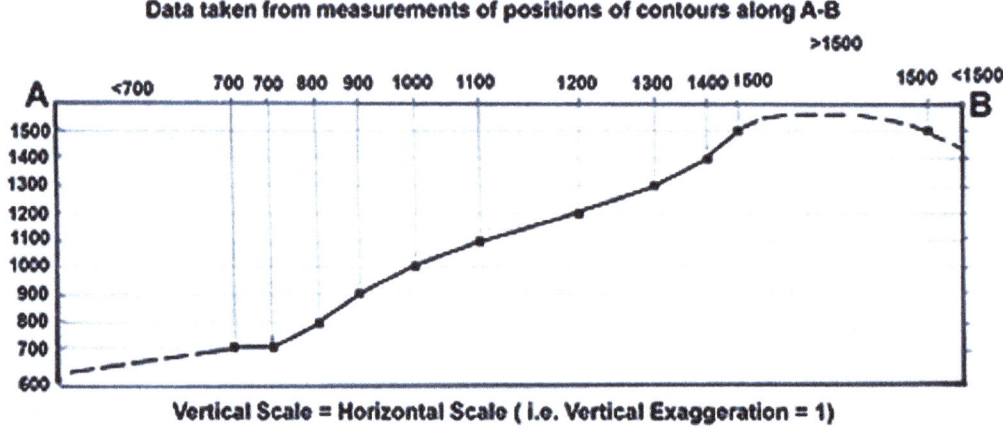

1. What feature lies between spot heights A and B?
   A generally uniform slope rising from A up a hill and then down to B

2. When drawing this cross-section, what allowances for reality must the drawer make?
   That real land surfaces are not like graphs and so an estimation of the surface must be made between intersections on the construction and also at each end.

3. What is the GRADIENT of the steepest slope along section A-B?

Remember that GRADIENT is the ratio of the vertical height to the distance covered horizontally on the map or "rise over run" or the tangent of the angle of that slope. On this map the distance along the actual slope, starting at the second 700 metre a.s.l. mark and going to the first 1500 metre a.s.l., is approximately 1050 metres and the difference in altitude is approximately (1500-700) is 800 metres so the gradient of the slope is about 800 divided by 1050 or 0.76. The angle with this tangent is about 37 degrees.

4. **If this is open grassland country and in good weather, what would be an estimate of the time it would take to walk (unburdened) from A to B?**
From Page 27 of the text, one finds that as a rule of thumb, for relatively open country, one should allow approximately:

   ONE HOUR for every 4 km. measured on the map;
   plus ONE HOUR for every 450 metres of vertical height climbed.

So, walking up this slope should take about 15 minutes for the 1050 metres walked along the ground PLUS just less than two hours for the 800 m vertical rise. In all about two hours. A decent hike!

CONCLUSIONS:

1. **What would be the (a) advantages and (b) disadvantages of using a paper topographical map over that of an electronic, hand-held Mapping Ap. or computer screen map?**
Advantages include the ready convenience of a paper map of large size without the small screen and possibilities of flat batteries. Paper maps and photos are limited in their scope and locality without the possibility of another quick reference.

2. **Why would an Earth Scientist use such a map? (consider pre- and post- field work activities)**
Paper maps can be used to measure distances and determine landforms more easily on the larger scale and they are useful in drawing up a blank base map of the same scale.

3. **What are some general conclusions about this mapping exercise?**
Depends upon the student's ability to apply the information. Hopefully that such old-fashioned ideas are still useful in an electronic age.

Some useful links are:

http://www.ga.gov.au/scientific-topics/national-location-information/topographic-maps-data/basics

https://education.usgs.gov/lessons/teachingtopomaps.html

https://education.usgs.gov/lessons/mapresources.html

https://store.usgs.gov/map-locator

https://serc.carleton.edu/mathyouneed/slope/index.html

http://www.fgmorph.com/fg_1_25.php

http://earthtoleigh.com/documents/worksheets/1.3%20Topographic%20Maps%20I.pdf
Good test worksheet on Topo. Maps (Open as pdf and Save)

http://earthtoleigh.com/documents/worksheets/1.3%20Topographic%20Maps%20II.pdf
Another good Topographical test/activity worksheet (Open as pdf and Save)

1.4 Other Activities

- **Using a Geological Pick** - at this early stage it would be a demonstration on a large rock specimen e.g. a friable shale or limestone. A large chunk of calcite is good if obtained because it cleaves nicely into crystals. Students (with discretion) may be given a pick and a smaller chunk of calcite and asked to GENTLY tap and further cleave their specimen. They can then use the hand lens to see the regular shape of calcite. Rock Salt may be another alternative.

    Online Video: The safe use of a geological hammer
    https://www.youtube.com/watch?v=VjQda-q1v1E

- **Internet/Cell Phone search** for FREE Earth and Environmental Science Apps to do with GPS location & navigation, use of maps, survival in the field, weather, minerals, rock types etc. Discuss advantages of Apps which do NOT rely on an internet/wifi connection. LIST the best from classroom search.

- **Discuss career opportunities and requirements in Earth and Environmental Science** - have a guest speaker if possible or a computer (Skype) link up.

    http://work.chron.com/10-earth-science-careers-10541.html

    https://www.environmentalscience.org/careers

    https://www.seek.com.au/environmental-science-jobs

    https://www.sciencebuddies.org/science-engineering-careers
    (a good site as it also shows that Earth and Environmental Science careers that may NOT need a college/university degree)

    http://www.bgs.ac.uk/vacancies/careers.htm?src=topNav

    https://www.prospects.ac.uk/careers-advice/what-can-i-do-with-my-degree/environmental-science

    https://environmentalsciencedegree.org/careers/

- **Research the life and work of a famous national Earth and Environmental Scientist** – a good library assignment to get students to use resources other than the Internet, although this can be also done using the Internet in class or at home.

- **Introduce students to Geological Maps** and stereo images if possible. Use/revise the use of normal topographical maps, a good overlap with Geography programs. If possible, have a set of local topographical maps and try some simple exercises such as:

  - identifying man-made features (buildings, roads, mines etc.)

  - identifying natural features (hills, cliffs, valleys, cols etc.)

  - use a piece of string, bootlace in an emergency, to measure the distance along a winding road or river.

  - Take any two points on the map, in hilly country if possible, and measure the distance between them using a ruler and the map scale. Discuss why this distance may be misleading. Determine how long it might take to walk this distance and what other factors are involved.

  - as an exercise, discuss with students what they would need to camp out for five days. Discuss the need for preparing for sudden changes in weather and suitable provision for extremes of hot and cold. What other environmental hazards, in the local wilderness area, would need to be considered?

1.5 Answers to Multichoice Questions

Q1.D Q2.B Q3.B Q4.C Q5.D Q6.A Q7.D Q8.B Q9.B Q10.D

1.6 Some Suggestions for the Review Questions

In many cases there may be several opinions for some of these questions and students will also provide a variety of answers and detail. Some possible additional suggestions or guides for the teacher are given below:

1. **Briefly discuss the role of the Earth and Environmental Scientist within today's society.**
   A good discussion! The role of the scientist has always been to look for answers to some of the world's great questions. Some of the more immediate problems such as disease, climate change, new technologies and over-population may be discussed. Certainly, it should be stressed that Science is on-going and a useful part of society. List some of the TYPES of science known to the students.

2. **What non-scientific factors also affect the role of the Earth and Environmental Scientist today?**
   Consider such issues as politics and popular opinion generated by the media, especially hysteria about global warming. Social factors, rights of native peoples and the debate about conservation versus exploitation of resources may be useful. Discuss barriers to scientific research.

3. **Why do research scientists develop a model when they need to answer major research questions?**
   Often the issues and relationships are difficult to see just by looking at the data. Scientists develop models to see which one will match the data. It is a sort of 'what happens if I do this' approach.

4. **What would be the major differences in thinking and methodology between an Earth and Environmental Scientist and a laboratory-based scientist, say a Chemist or Physicist?**
   It would be a generalization to say that Earth and Environmental scientists do field work and the others are only in a laboratory. Earth and Environmental Scientists certainly do a considerable amount of work in the field but field work is also undertaken in some occasions by the others, have students suggest some examples. Earth and Environmental Scientists also work in laboratories, some exclusively, preparing field work and then analysing field data. The philosophies involved in the scientific method are the same for all but the emphasis on where data is gathered may be different. Also, all sciences are global in outlook.

5. **Earth and Environmental Science can be an adventurous science. Is this true? Discuss.**
   Certainly, the author has found it to be! All science is an adventure because there is the constant search for answers with the generation of new questions, but field scientists also must venture into some very adventurous places, often with dangerous terrain, climate, wildlife and sometimes local inhabitants. Antarctica and the Amazon jungle were two of the author's most interesting trips.

6. **Define each of the following terms:**

   a. **Isolated system** - a system that does not interact with its surroundings; that is, its total energy and mass stay constant.
   b. **vertebrate palaeontology** - study of animals which have a bony spinal column.
   c. **Meteorology** - the study of the atmosphere and weather patterns.
   d. **scientific model** - a scientific model is a representation of a particular phenomenon in the world using something else to represent it, making it easier to understand.

7. **Distinguish between the terms validity and reliability. How do they relate to accuracy in using data?**
   The validity of any observation, experiment or gathering of data, is how fairly or honestly the results tests the hypothesis in question whereas the reliability of any measurement is its ability to be repeated and show consistency in its data and

calculations. Any problem involving either or both of these factors decreases the accuracy of any measurement because there would be doubt as to whether or not the measurement is really about the question being answered and if values of measurement are very random and may involve large errors.

8. **Research the institutions in your local area that offer courses in Earth and Environmental Science. What would be the pathway in your area in obtaining a position as an Earth and Environmental Scientist.**
   This will vary depending upon local applications of Earth and Environmental Science and any local institutions. Formal pathways (e.g. High School to College or University etc.) should also be compared with the usefulness of informal interests and hobbies in the Earth and Environmental Science which often can be lucrative as well.

9. **What basic equipment should be carried by a field scientist on a two-day traverse in regard to** (A few ideas only):

   a) **sampling geological data** - the usual compass and map (also GPS unit), pick, specimen bags, hand lens, chisel, notepad/pencil (ink runs in wet weather!)
   b) **basic living** - water & food, shelter (light tent/army hootchie), good clothing, hat and boots, cooking utensils as needed, pack, wet/cold weather jacket, insect/sunburn cream.
   c) **safety**- appropriate clothing, small first-aid kit, cell phone/radio (if reception possible), aluminized space blanket, extra water. Try not to travel alone and ALWAYS have an emergency plan (route out or shelter) and leave details of timings and route with some local authority.
   d) **navigation** – maps and compass. Cell phones and GPS units are good but have no alternatives if they fail. Research the area first and have a general idea of the surface.

10. **Soils and vegetation can often indicate local geology. Discuss the role of local animals in indicating geology (an interesting question which might need some extra research).**
    Research and some examples from experience needed here e.g. animals take the easy route so follow trails through difficult country; some species of vegetation only live in certain soils; in limestone (Karst) country, lone big trees often grow near moist sinkholes.

1.7 Reading List

Carson, Rachel. (1962) *Silent Spring*. Boston: Houghton Mifflin. Reprinted 2002 by Mariner Books, ISBN 0-618-24906-0

Gohau, G. (1990). *A history of geology*. Revised and translated by Albert V. Carozzi and Marguerite Carozzi. New Brunswick: Rutgers University Press. ISBN 978-0-8135-1666-0.
Harbaugh, J.W. Geology. 2015. *Encyclopaedia Britannica*.
http://www.britannica.com/science/geology

Iowa State University. (2015). Glossary of Geologic Terms. *Dept. of Geological and Atmospheric Science.*
http://www.ge-at.iastate.edu/glossary-of-geologic-terms/

Kent Educational, 2015. *Careers in Geology.*
http://www.personal.kent.edu/~cschweit/Stark/whygeologymajor.htm

McLelland, C.V (Edit.).(2006) *The Nature of Science and the Scientific Method.* Geological Society of America. http://www.geosociety.org/educate/NatureScience.pdf

Mathez, A. (2000). *Earth: Inside and Out.* New York: New Press, American Museum of Natural History. Excerpts at:
http://www.amnh.org/education/resources/rfl/web/essaybooks/earth/

Ohio University. (2015). Careers in Geology. *College of Arts and Sciences.*
https://www.ohio.edu/cas/geology/careers/about.cfm

Scott, P.T. (2016). Exploration Science. Brisbane: Felix Publishing. ISBN 978-0-9946432-8-5

Study.com. (2015). *Chapter 26: What is Geology?*
http://study.com/academy/lesson/what-is-geology-definition-history-facts-topics.html

What is Environmental Science (2018) https://www.environmentalscience.org/

Photomicrographs at:

http://www.idahogeology.org/PDF/Digital_Data_(D)/Digital_Analytical_Data_(DAD)/all_Challis.pdf

https://lifeinplanelight.wordpress.com/2011/03/01/photomicrograph-tuesdays-k-feldspar/

http://pcwww.liv.ac.uk/~hiatus/resources/igneous.htm

https://www.earth.ox.ac.uk/~oesis/micro/medium/anorthosite_pm18-35.jpg

http://www.esta-uk.net/geooer/wex%20photomicrographs%20-%20igneous.htm

About Feldspar at:

http://geology.com/minerals/feldspar.shtml

http://www.ima-na.org/?page=what_is_feldspar

Chapter 2: The Blue Planet

2.1 Theme

The Earth is a planet orbiting a star, our Sun as part of the Solar System which is an open system. This motion, the Earth's rotation and the tilt of its axis are major influences on the various environmental spheres of the planet.

2.2 Rationale

An understanding of the nature of planet Earth assists in understanding of how the processes within each of the earth's four spheres operate. The tilt of the Earth explains the seasons which in turn affect the weather and the range and interaction of living things on the surface. The gravitational influence of the Sun and the Moon affects the tides in the oceans.

2.3 Notes on the Practical Work (see PRACTICAL MANUAL 1.)

Experiment: 2.1: The Rotation of the Earth

The movement of shadows because of the Sun's apparent motion across the sky has been a matter of study and application of Humankind for many thousands of years. Many of the world's great astronomical structures, such as temples in ancient Egypt, Stonehenge in England, the Intihuatana of Machu Picchu and the great observatories of India have all been used to accurately plot this passage for religious and farming calendars.

In theory the shadow of any upright structure should show a complete rotational angle of 360 degrees in approximately 24 hours (i.e. one complete rotation of the Earth) and it does if one times the rotation over one day from one exact time (say noon) to the next. However, in small scale measurements such as this experiment, one would find that the shadow speed will vary from time of day and latitude. See "Apparent movement of the Sun" at:

http://energy-models.com/earth-and-sun

Also shadows of big things move faster (at the tip) than shadows of small things, and shadows grow or shrink very fast at sunset and sunrise.

A great website for angles of the Sun at different locations and times is at:

https://www.timeanddate.com/sun/australia/brisbane?month=3

If one takes the time difference between consecutive sunrises it should show approximately the time for one rotation e.g. on 27th April 2018 at Brisbane, Australia, the Sun rose at 6.10 am and on the 26th it rose at 6.11 am. This of, course changes from season to season with later sunrises in winter. This time difference of 24 hours (and one minute) dividing into a complete rotation of 360^0 gives a figure of 15^0 per hour. Because of the factors mentioned above, students will be very lucky if they obtain this value, but the activity is worth doing to show the basic ideas of the use of a shadow stick in time and for navigation.

The first part of the experiments requires the measurement of the length of shadows and the subsequent angles they subtend at different times. Ideally this should be done regularly over the whole period of sunlight but this is not practical. With the constraint of timetables and with classes early in the morning and late in the day (or night) it would difficult to do this. A compromise would be to measure the shadow parameters around noon from (say) about 11.00 am to 1.00 pm at 15-minute intervals. This may need some time out of class.

Another good Sun calculator is at:

https://www.suncalc.org/#/-27.4698,153.0251,14/2018.04.27/11:02/1/0

This calculator shows the Sun's shadow for a one metre shadow stick and the times can be changed, giving a virtual version of the first part of this experiment. For example, at Brisbane at Latitude South 27°28'11.17" (other locations in the US and UK are available) on the 27th April, 2018, the shadow of the one metre stick were:

LOCAL TIME	SHADOW LENGTH	SUN ALTITUDE	SUN AZIMUTH	AZIMUTH CHANGE/HOUR
10.00 am	1.13 metres	41.49 degrees	35.21	18.75° per hour
10.15	1.03	43.30	30.93	
10.30	1.00	44.90	26.37	
10.45	0.96	46.26	21.54	
11.00	0.92	47.32	16.46	
11.15	0.90	48.12	11.17	21.74° per hour
11.30	0.88	48.60	5.74	
11.45	0.88	48.77	0.23	
12.00 noon	0.88	48.62	354.72	
12.15 pm	0.90	48.16	349.28	20.68° per hour
12.30	0.92	47.38	343.98	
12.45	0.95	46.32	338.89	
1.00	1.00	44.99	334.04	
1.15	1.06	43.41	327.46	16.62° per hour
1.30	1.13	41.61	325.16	
1.45	1.21	39.61	321.15	
2.00	1.31	37.44	317.42	

The SUN ALTITUDE is the angle that the Sun makes with the ground subtended by the shadow stick of one metre.

The SUN AZIMUTH is the geographical bearing of the Sun (e.g. North is 360°) at that time.

From this data one can see that:

- Solar noon, when the Sun's shadow is shortest is between 11.30 and 12.00 noon (more precisely by Sun Altitude near 11.45 when the angle is at its highest. When a sailor goes to use a sextant, he/she will use it around noon to find true Solar noon by finding the time when the angle of the Sun is highest (Graph 1):

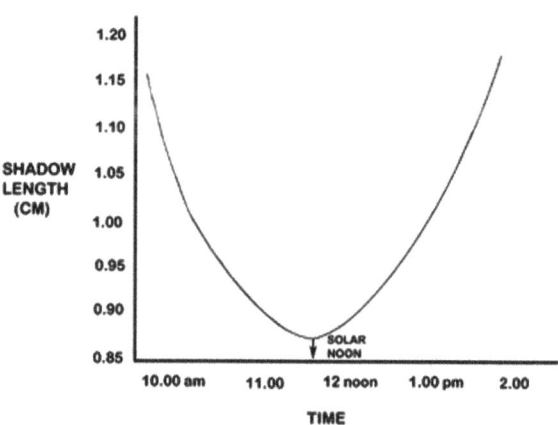

- at Solar noon, the Sun is also at its maximum Azimuth (in theory North but this varies a little due to the Earth's tilt).

- the angle swept out, change in Azimuth, by the shadow is smaller when the time is away from noon and faster around noon. A graph of the cumulative (addition) of the azimuth changes against time and extrapolated for 12 hours may give a better reading but it will still not be 15 degrees per hour. This will have to be seen over one complete 360-degree rotation.

TIME	CUM. AZIMUTH CHANGE
10.15 am	4.3 degrees
10.30	8.8
10.45	10.7
11.00	15.7
11.15	30.5
11.30	25.8
11.45	31.8
12.00 NOON	37.7
12.15 pm	42.8
12.30	48.2
12.45	53.6
1.00	58.7

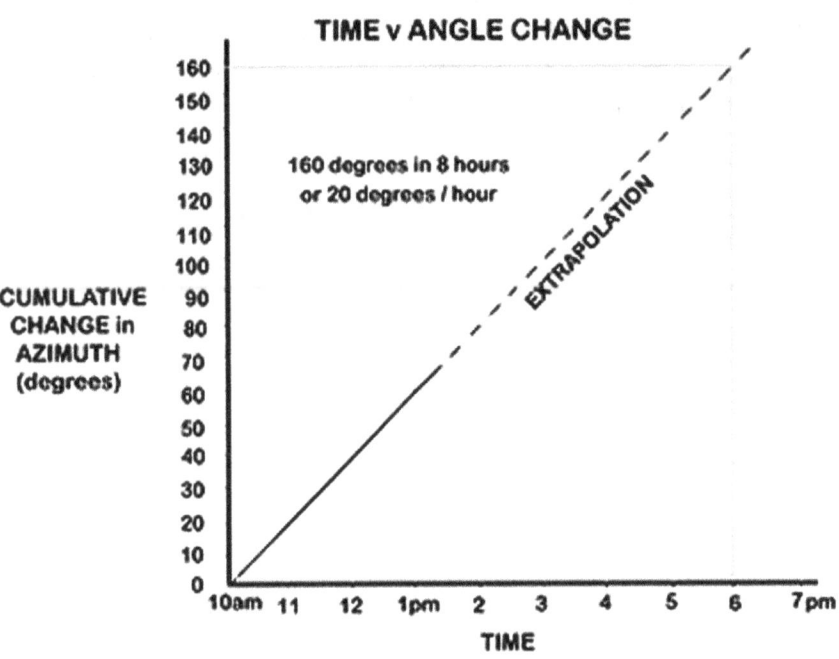

CONCLUSIONS:

1. **Describe the shape of GRAPH 1.**
   It will give a saddle-like curve with a dip at the lowest point at solar noon.

2. **What was the time for the shortest shadow (Solar Noon) from this graph?**
   This depends upon the location but it will be at the lowest point.

3. **Was this value exactly 12.00 noon? If not why was it different?**
   No. It varies depending upon the locality and the season.

4. **What was the shape of GRAPH 2? What does it say about the rotation of the Earth?**
   It is a straight line, incline, showing that the rate of rotation over extended time was uniform.

5. **If the shape of GRAPH 2 was extrapolated, what was the angle or time for one rotation of the Earth? Calculate the percentage error for this calculation.**
   This depends on the graph and the extrapolation; the percentage error is the difference from 15 degrees/hour divided by the rate obtained then multiplied by 100. It may be quite high!

6. **How could this measurement (step 9) be made more accurate?**
   Use a longer shadow stick over a longer period of daylight near the Equinox for that location.

RESEARCH: (Optional)

Use the Internet to find out about:

1. **Sidereal time and Earth time.**

    See:

    https://www.howitworksdaily.com/how-is-time-measured/

    https://nrich.maths.org/6070
    (a good history of time measurement)

    http://www.npl.co.uk/educate-explore/what-is-time/the-world-time-system

    http://www.astronoo.com/en/articles/measuring-time.html

2. **Instruments used by Tycho Brahe** – this is included because this astronomer was such a fascinating character, read about his personal life, but also because he used large constructions to measure aspects about the Earth's rotation and position in space. See:

https://www2.hao.ucar.edu/Education/FamousSolarPhysicists/tycho-brahes-observations-instruments

http://ircamera.as.arizona.edu/NatSci102/NatSci102/lectures/tycho.htm

http://www.renaessanceforum.dk/12_2017/12_christianson_brahe.pdf

http://www-groups.dcs.st-and.ac.uk/%7Ehistory/Biographies/Brahe.html
(His personal life)

Experiment: 2.2: Tides

This a paper exercise using a tide chart. These are very useful for a range of fishing, boating and scientific activities. Tides vary depending upon the relative positions of the Sun and the Moon but they also vary due to the coastal locality because of such factors as barrier islands and the shape of the coastline.

The exercise can be tiresome but worth the effort. Students will have to copy the graph (electronically -cut-and-paste) or copied out at an appropriate scale onto paper as required. They will have to make their own vertical scale (say 0.0 to 2.0 m).

The final graph will look like the typical tide chart with rounded peaks - forgive the abbreviated version here:

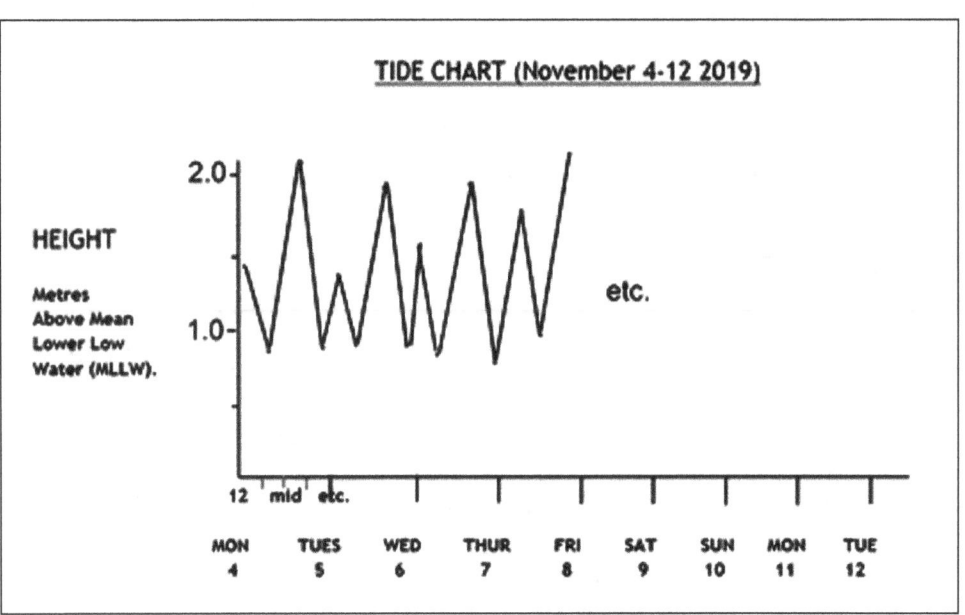

CONCLUSIONS:

1. **What does the graph show about tides?**
   That they are cyclic with this pattern showing approximately two highs and two lows each day (semi-diurnal).

2. **When did the highest and lowest tides occur? Why?**
   They seem to have high tides around Noon when the gravitational attraction of the Sun is highest and lows near midnight. This graph does not show the position of the Moon which is the main influence.

3. **What other factors would affect the time of the various tides in this location?**
   These include the shape of the coastline, proximity of local islands, ocean currents and winds.

4. **Why are a knowledge of tides useful?**
   Apart from recreational activities such as fishing, a knowledge of the local tides is useful when studying coastal environments, especially in places where there is a great, and dangerous, tidal surge. Also, if the research requires the use of boats, a knowledge of the tides within boating channels is essential.

5. **What type of tidal pattern does this graph show?**
   Semi-diurnal.

RESEARCH: (Optional) Local tide charts
Locate a local time chart for the current month. Does it agree with findings from this experiment? This is useful for coastal schools and those on tidal rivers. Inland students

with little knowledge of tidal patterns should study the closest coastal port or one with a bad reputation for coastal bars and waterways.

See: https://manoa.hawaii.edu/exploringourfluidearth/physical/tides/tide-patterns-and-currents

http://www.bom.gov.au/oceanography/projects/ntc/tide_tables.shtml
(Tide charts for places in Australia)

https://www.tide-forecast.com/countries
(Tide charts for international locations)

2.4 Other Activities

- Construct a sundial, either as a temporary device or a permanent one at home. See:

    http://hackaweek.com/hacks/?p=915

    http://www.skyandtelescope.com/astronomy-resources/how-to-make-a-sundial/

    http://hilaroad.com/camp/projects/sundial/sundial_calculator/sundial_calculator.htm with angle calculator

    http://www.sundials.co.uk/projects.htm

    http://www.instructables.com/id/15-minute-paper-craft-sundial/ and its calculator at http://analemmatic.sourceforge.net/cgi-bin/papercraft.pl

- Use the Solar Angle calculator website (from Expt. 2.1) and several different locations which are due north and south for noon on a given day to find other estimations of the Earth's circumference.

- If near the coast, go on a field trip to examine the life found in tidal pools but make sure that this is done safely and at low tide. Know the time for an incoming tide. See:

    https://www.qld.gov.au/environment/coasts-waterways/marine-habitats/rocky-shore

    https://parks.des.qld.gov.au/experiences/reef-activities/reef_walking.html

    https://montereybay.noaa.gov/visitor/TidePool/welcome.html

- Other tidal activities at:

    https://www.lpi.usra.edu/education/workshops/phasesSeasons/resources/tides/

2.5 Answers to Multichoice Questions

Q1.C Q2.A Q3.B Q4.B Q5.D Q6.B Q7.C Q8.A Q9.D Q10.D

2.6 Some Suggestions for the Review Questions

1. Explain the differences between each of the following:

    a. **a star and a planet** - stars emit energy, planets orbit stars.
    b. **orbital period and rotational period** - time to make one complete motion around the Sun and the time to spin on its axis once.
    c. **vernal and autumnal equinoxes** - Vernal equinox is when the Sun is overhead in the spring and the autumnal equinox occurs in autumn <u>in the Northern Hemisphere</u>. In the Southern Hemisphere the times are reversed (i.e. the vernal equinox occurs in March-June)
    d. **spring and neap tides** - Spring Tides, when the water 'springs up', occur when the moon is full or new, the gravitational pull of the moon and sun are combined and tides are higher than usual. Neap tides are when the Moon and Sun's gravity are at right angles and give lower tides.
    e. **speed and acceleration** - speed is the change in distance per time and acceleration is the change in velocity (speed in a given direction) per time.
    f. **diurnal and semi-diurnal tides** - diurnal tides one high and low per day but semi-diurnal have two cycles per day.
    g. **habitats and ecosystems** - habitats are the natural homes of particular animals, plants, or other organisms whereas an ecosystem is the broader community of many habitats within an environment.
    h. **hydrosphere and geosphere** - the hydrosphere concerns the waters of the Earth whereas the geosphere concerns the land.

2. **Explain the meaning of the term accretion. Why was it important in the development of planet Earth? What forces were involved in this process?**
Accretion is the joining of items which have collided and now stick together. It is thought that the planets such as Earth formed when debris in the early disk of the young solar system collided and stuck together.

3. **The Sun is a burning mass of gas. Discuss this statement.**
False. The Sun is a fusion nuclear body in which hydrogen plasma (super-heated charged gas) nuclei combine together to form larger atoms such as helium. There is no free oxygen gas (O_2) in the Sun to combine with any fuel as in burning.

4. **Research the dates which mark the beginning of the four seasons in your locality.**
In Australia (southern hemisphere) the seasons begin on the first day of the calendar month, with summer from December 1 to the end of February, autumn from March to May, winter from June to August, and spring from September to November. In the Northern Hemisphere the seasons are reversed i.e. spring runs

from March 1 to May 31, summer from June 1 to August 31, autumn (fall) from September 1 to November 30, and winter from December 1 to February 28

5. **Discuss why a knowledge of local weather and tides would be important to an Environmental Earth Scientist.**
Field work is the most important way of gathering data. Climatic seasons often determine when specific organisms are available. Tidal information assists in the access to organisms living between the tide zones and also indicates any dangers from sudden tidal surges or boating access to channels. The weather always is a critical factor in field research due to the possibility of extreme conditions but also in comfort of exploration and access to organisms.

6. **Freshwater suitable for drinking only makes up only about 2% of the hydrosphere. With an increasing world population and global warming, discuss methods which could be used to increase freshwater supplies to a community which has lost its original source. Discuss several community types e.g. coastal, inland, mountain.**
This could be a very long discussion. It may include engineering projects such as dams and pipelines to channel water from where it is in good supply to where it is not. It might involve desalination projects to convert seawater into freshwater etc.

7. **What is the Gaia hypothesis? Discuss this with some examples.**
Proposed by James Lovelock in the 1970's this proposes that living organisms interact with their inorganic surroundings on Earth as a self-regulating, complex system that maintains and perpetuate the conditions for life on the planet e.g. excess carbon dioxide is regulated by the use of photosynthesis in green plants. Still some doubt about this hypothesis.

8. **Research the Butterfly Effect. What does it have to do with the four spheres of the Earth?**
A metaphorical (made-up example) coined by Edward Lorenz for the idea that small causes, e.g. flapping of a butterfly's wings, may have large effects elsewhere, such as causing a tornado. It could be used to suggest that small events could accumulate into larger events elsewhere within the Earth's spheres such as in the atmosphere, small currents to tornados, minor soil flow to landslides, small ripples to large ocean waves. An interesting idea.

9. **What type of system would each separate Earth sphere be classified under? Explain.**
Earth spheres are all open systems because they lose both matter and energy to each other and indirectly off the planet.

10. **Research an indigenous legend about the Sun, day and night or time from a culture with which you identify (e.g. Celtic, Anglo-Saxon, Aboriginal, Indian, Chinese, Japanese, Middle Eastern, other European etc. etc.) Compare this with others in your group, this could be used as a display.**
There are many great ideas coming from ancient (and some modern) indigenous peoples about the origin and processes of the Earth. These are all of value because they were useful in explaining why things happen, allowed for predictability and

also were used as teaching stories for the next generation. A good project would be for students, especially those who know their roots to prepare one legend to be discussed or displayed in the class.

2.7 Reading List

Campbell, Neil A.; Brad Williamson; Robin J. Heyden (2006). *Biology: Exploring Life. Boston, Massachusetts: Pearson Prentice Hall.* ISBN 0-13-250882-6.

Cartwright, D. E. (1999). Tides: A Scientific History. Cambridge, UK: Cambridge University Press. 292 pp. ISBN 0 521 62145 3

Cattermole, Peter & Moore, Patrick *(1985). The Story of the Earth. Cambridge: Cambridge University Press.* ISBN 978-0-521-26292-7

Cockell, Charles (2008). *An Introduction to the Earth-Life System.* Cambridge University Press.

Gillard, A. 1969. *On Terminology Of Biosphere And Ecosphere.* Nature 223:500-501.

Fothergill, A. (author) and Attenborough, D. (forward) (2007). *Planet Earth as you've Never Seen it Before.* University of California Press. 312 pp. SBN: 9780520250543

Izidoro, A., Haghighipour, N., Winter, O. C. & Tsuchida, M. (2014). Terrestrial Planet Formation in a Protoplanetary Disk with a Local Mass Depletion: A Successful Scenario for the Formation of Mars. The Astrophysical Journal 782 (1): 31, (20 pp.). arXiv:1312.3959. Bibcode:2014ApJ782231I. doi:10.1088/0004-637X/782/1/31.

Jacobson, Michael, RJ Charlson, H Rodhe, GH Orians: *(2000). Earth System Science, From Biogeochemical Cycles to Global Changes (2nd ed.).* London: Elsevier Academic Press. ISBN 978-0123793706.

Lang, Kenneth R. (2011), *The Cambridge Guide to the Solar System*, 2nd ed., Cambridge University Press.

Malville, J. M., Thomson, H., and Ziegler, G. (2006), *The Sun Temple of Llactapata and the Ceremonial Neighborhood of Machu Picchu*, in Bostwick T. & Bates, B. (eds.), *Viewing the Sky Through Past and Present Cultures.* City of Phoenix Parks, Recreation and Library, Phoenix, pp.327-339.

Moore, Patrick & Rees, Robin. (2014). *Patrick Moore's Data Book of Astronomy.* Cambridge, U.K. Cambridge University Press. ISBN 1139495224.

Scott, P.T. (2018). Beyond Planet Earth. 2nd edition. Brisbane: Felix Publishing. ISBN: 978-0-9946432-4-7

Some Useful Free Astronomy Apps:

Tides near Me (free but simple app for giving local tides in many places of the World).

Tide Table (is another free and simple tide data app with useful data).

My Tide Times (Tide Tables, Forecasts & Tides! - a free app which is very Comprehensive).

Chapter 3: Exploring the Rocky Planet

3.1 Theme

This chapter is an introduction to the Geosphere and shows how the science of Geology developed as people, mostly practical men and women such as canal builders and fossil collectors gradually developed an understanding of how different landforms and rocks came about and the relationships between them.

3.2 Rationale

An understanding of how layers of rocks and igneous intrusions formed and are related is vital to putting together an idea of how parts of the Geosphere were formed. This geological history often comes about because of many separated ideas being pieced together to form a broader picture. From here, the past events of the region and probably some predictions about the future may be made.

3.3 Notes on the Practical Work (see PRACTICAL MANUAL 1.)

<u>Experiment: 3.1</u>: Correlation Exercise

This is a simple paper exercise to show the important principle of correlation. Whilst this exercise uses layers of sedimentary rock, it should be pointed out that correlation can be done on a national and international scale with fossils, rock structures and radiometric ages.

Photocopies or electronic copies of the 'stratigraphic columns' could be made, even on a larger scale to be coloured and then pasted (or scanned) into the student report. If the school has access to drill cores then these would be excellent to show the students how they are logged. Artificial 'drill cores' could be made from lengths of coloured and textured wood or cardboard rods or rolls and students could log them.

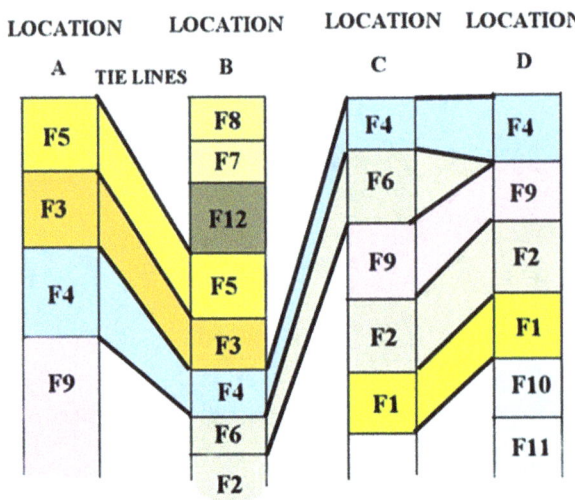

Note that to keep the diagram uncomplicated the beds which would lens out between columns have not been continued.

The sequence of these beds from oldest to youngest would be:

F11 (oldest) F10 F1 F2 F9 F6 F4 F3 F5 F12 F7 F8 (Youngest)

CONCLUSIONS:

1. **What errors could occur in using this method to determine relative age of beds?**
   It does not allow for any over-turned beds nor faulting of the sedimentary sequence. Also, there is only limited correlation between cores B and C.

2. **How important is the accuracy of identification of the fossil specimen in this process?**
   Very, as it allows for better correlation. Some fossil specimens may have only slight differences in appearance but be of different time frames.

3. **Give an hypothesis for why Fossil F6 is missing from columns A and D.**
   It may have been eroded away from these locations or more likely it lensed out between the core locations having only a small area of deposition.

4. **This exercise has fossils in all rock units. This is not common as fossils are very difficult to find. Give some reasons why fossils may not be found in rock units at all.**
   There are many reasons and is most often the case. Reasons include: Life forms did not exist in the region; their remains were removed by erosion and land disturbance; or conditions were not suitable for preservation.

Experiment: 3.2 Mapping Inclined Beds

Sedimentary rocks, basalt lava flows and intruded sills would be horizontal if there were not internal forces in the Earth which lift and tilt them. The mapping of tilted beds is perhaps one of the simplest field activities, especially if the whole sequence of normally horizontal beds is all tilted or inclined in the same direction. Traditionally a field geologist would establish a route (or traverse) which would go in the direction in which the beds are tilted, this is the dip direction, and so measure the angle of tilt, angle of dip, on flat bedding planes. Sometimes it may not be convenient to go in the dip direction but across it. In this case, the geologist would measure angles which may be on flat surfaces which are oblique to the true dip, called the apparent dip. A good field geologist would still measure the true dip on these oblique surfaces by finding the true dip direction which would be at 90^0 to the strike of the beds, the geographical direction in which the beds cut the surface, or in a difficult situation where the strike also cannot be seen, the geologist may trickle some water down the oblique incline and see which way the water runs; usually in the true dip direction. Students should revise how the Brunton Compass is used to find general direction, e.g. dip direction, strike, as well as the angle of dip.

See the video from the textbook on the use of the Brunton Pocket Transit at:

Online Video: How to use a Brunton Pocket Transit.
https://www.youtube.com/watch?v=BrTekM0DRQk

PART A: Mapping Dipping Beds in the Dip Direction

This is a relatively straight forward exercise but notice that data cannot be displayed with any certainty below the bottom edge (at B) for the conglomerate. For simplicity, in all of the following exercises, vertical scale = horizontal scale.

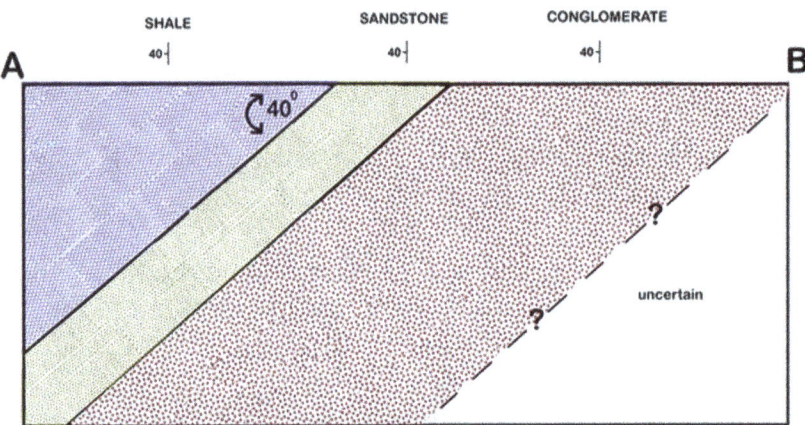

PART B: Mapping Dipping Beds Not in the Dip Direction

With a true dip of $40°$ and a direction angle of 200, the table gives an apparent dip of $38°$. This is not much, but on a larger scale, it could be significant.

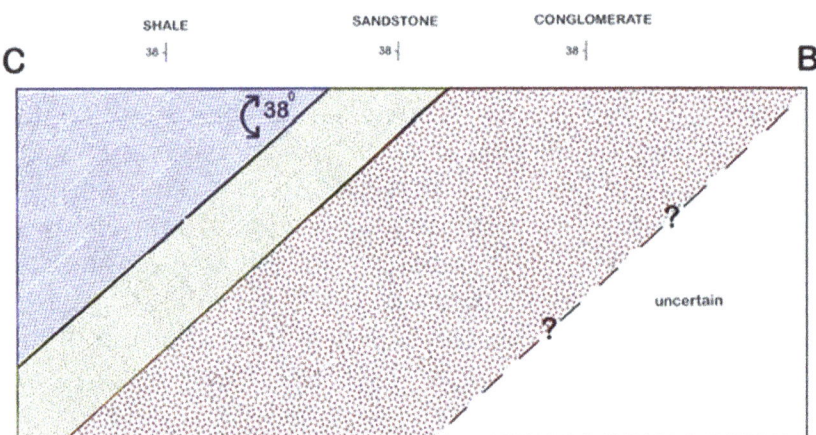

CONCLUSIONS:

1. **Why would a geologist NOT walk along the Dip Direction to measure true Angle of Dip?**
   Any numbers of reasons such as there were only limited exposures of the beds, the surface was too rough or weathered, the land may be covered in scrub or it may be fenced off.

2. **What other geological or environmental situations cause difficulty in measuring the true dip of beds?**
   These include the limited exposure of good bedding planes, excessive weathering of the surface and vegetation cover, just to name a few.

3. **Why is dip direction a better parameter than strike when discussing the orientation of beds?**
   It is more practical to follow one direction along a traverse than to constantly trying to determine the perpendicular orientation of strike to the plane on which the angle of dip is to be measured. Some beds which have good strike but have suffered some deformation in respect to overlying and underlying beds may need the use of the strike direction.

4. **In Part A, at what depth would a drill core meet the top of the conglomerate if the drill was started on the surface along the traverse at the contact of the shale and sandstone (Hint: use the cross-section and its scale)?**
   Dropping a vertical line straight down from the shale/sandstone boundary until it meets the top of the conglomerate would be about 300 metres.

Experiment 3.3: Mapping Faulted Beds

This exercise shows a fault with its fault plane dipping towards B at 70° and the sedimentary strata dipping towards A at 40°. If there is no dip shown on any fault, then it is assumed to be vertical. In fault mapping questions, the fault is the main feature as it is assumed that it cuts right through the section from top to bottom, so it is put in first with a solid straight line and the usual FF symbol.

Then the bedding planes are drawn in remembering not to pass them through the fault plane:

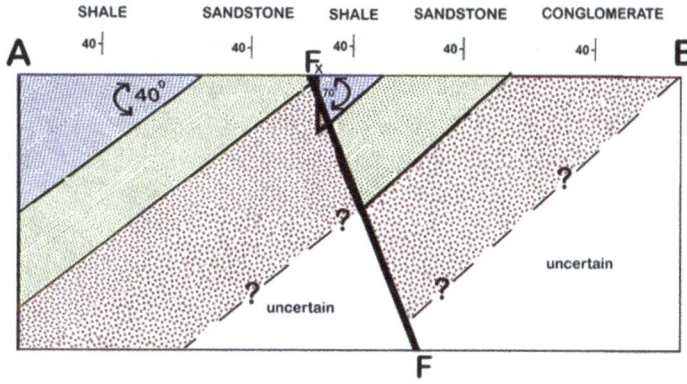

CONCLUSIONS:

1. **What type of fault is this? Which side of the fault has moved up or down?**
   This is a normal or tensional fault with the right-hand side towards B dropping down.

2. **Which side of the fault is the (a) hanging wall (b) the footwall?**
   The hanging wall is on the right of the fault plane and the footwall is to the left.

3. **What is the (a) heave and (b) throw of the fault?**
   Using the top of the sandstone bed as a marker, and by measurement, the heave (horizontal displacement) is about 1600 metres using the horizontal scale on the map, not the vertical scale on the cross-section. Now the vertical scale must be used and a line must be drawn projecting the top of the sandstone bed down to the left, through the fault plane so that it is now vertically below the same top of the sandstone on the left of the fault. Measuring the vertical displacement down from this point, the throw is about 400 metres.

4. **What type of Earth forces would cause such a fault?**
   This fault is caused by tension i.e. forces pulling apart at the fault.

5. **If a drill was put down vertically at X, at what depth would it strike the top of the conglomerate bed?**
   Projecting a line straight down from X until it meets the top of the conglomerate it will be reached about 100 metres below the surface.

Experiment 3.4: Mapping Folded Beds

With relatively slow compressional force, sedimentary strata can be folded into anticlines (upfolds), synclines (downfolds) or a combination of these. On the surface, the geologist will walk across repetitions of the same beds but dip directions will be opposite. In symmetrical folds, the angles of dip will be the same in these different dip directions but in asymmetrical folds the angles of dip will be different. Fold systems can be very complicated and difficult to map if they are also plunging (i.e. the whole system is made to dip other directions as well), as recumbent folds or dipping in all directions to produce domes or basins.

In this simple exercise, the fold axes should be lightly pencilled in first so that the sedimentary beds can be folded around them. Remember to round the curve of the folds because folds are rarely as sharp as drawing two lines together at an angle. Also have the symbols for the rocks orientated such that any linear symbol is parallel to the bedding plane (especially for the limestone and the shale).

Also remember to curve the bedding planes around at the fold axis. Do not run the bedding planes straight through them. If the fold is asymmetrical, one will have to find the curve of best fit, to curve the bedding around the fold axis. Remember that these cross-sections are meant to be a representation of the subsurface geology not a simple geometrical exercise.

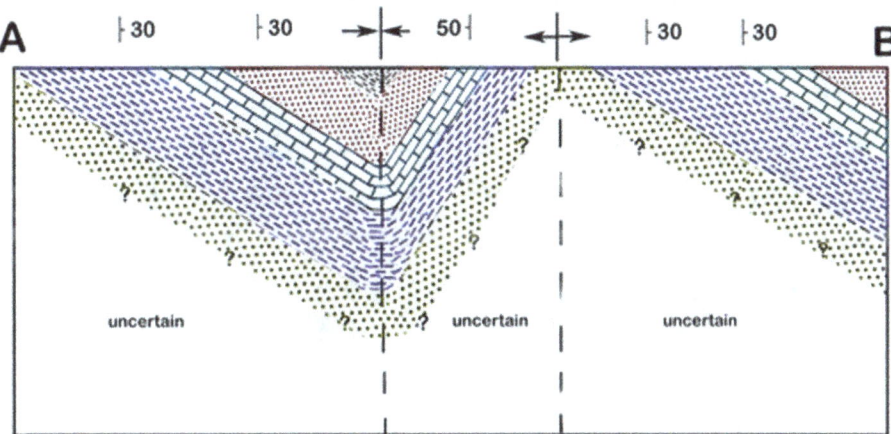

CONCLUSIONS:

1. **What type of fold is (a) closer to A and (b) closer to B?**
   There is a syncline at left and an anticline at right. Both are asymmetrical folds.

2. **What type of Earth forces would cause such folding?**
   Compressional forces from both sides.

3. **Which sedimentary beds were deposited (a) first and (b) last?**
   The coarse sandstone was deposited first and is overlain by the shale, limestone, fine mudstone and lastly the mudstone.

4. **When did the folding occur in relationship to the deposition of the beds?**
   The folding occurred after all of the beds had been deposited.

Experiment 3.5: Mapping Igneous Intrusions

The key to mapping igneous intrusions is that they usually occur after any other rock formation and intrude through it. Batholiths and stocks come up from deep below and usually have a broader base than their top. Dykes come up through cracks, cutting across strata (i.e. are discordant) whereas sills squeeze in between bedding planes and are parallel (concordant) with them. Unless a dip angle is given on any dyke, it is assumed that it is vertical.

The actual shape and extent of the intrusion below the surface is usually unknown unless drill cores have been obtained from around the outcrop margins. Also, the metamorphic aureole surrounding the intrusion could also vary in size and composition. In this exercise, there would be no surface indication of the metamorphic action below the surface; here the sandstone would become quartzite (at left) and the limestone would become marble (on the right).

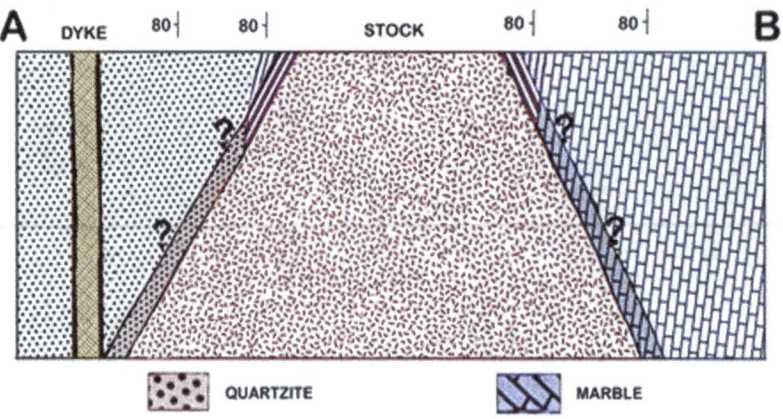

CONCLUSIONS:

1. **What types of igneous intrusions are seen on this map? Explain the reasons for your answer.**
   The narrow intrusion cutting across the surface as a thin sheet would be a dyke as it cuts across the dipping rocks. The large, almost oval intrusion would most likely be a stock because of its small size. Batholiths are usually much larger, being tens of kilometres across or more.

2. **What is hornfels? Why does it ring the central body?**
   Hornfels is the metamorphic rock formed by the strong heating of shale which has been intruded. The ring of hornfels represents the metamorphic aureole formed by the heat of the intrusion.

3. **Why is there no ring, or major edge along the dyke?**
   There is but very small and difficult to show at this scale. It may only be a few centimetres thick.

4. **What is the relationship between the rocks granite and aplite?**
   Aplite is the fine-grained version of granite so there is the suggestion that the two intrusions came about the same time and from the same magma chamber but this is uncertain.

5. **Is there any indication which intrusion came first?**
   Not clear (see above). There would have to be some drill cores taken between the stock and the dyke to see whether or not the dyke cuts the stock at depth (a good possibility also).

RESEARCH: (Optional)

1. **Types of Igneous Intrusions:**

http://www.classzone.com/vpg_ebooks/ml_earthscience_na/accessibility/ml_earthscience_na/page_125.pdf

https://volcanohotspot.wordpress.com/2017/07/06/batholith-lopolith-sill-or-dike-intrusions-2/comment-page-1/

2. Commercial uses of Igneous Intrusions

http://www.actforlibraries.org/what-are-igneous-rocks-used-for/

https://geology.com/rocks/pegmatite.shtml

https://pdfs.semanticscholar.org/095b/6710f69f492da9dc8d4f8decc1265056681a.pdf

Experiment 3.6: Mapping Unconformities

Unconformities are common in the geological record and they represent an hiatus or break in time when an existing set of rocks are uplifted and eroded. After some time, new sedimentation may occur giving the unconformity. Angular unconformities, disconformities and nonconformities are usually very obvious as they have a change in surface shape from one set of deposition (or intrusion) to the next as well as a zone of weathering between them. Paraconformities in which the deposition stopped for a while, was eroded to a flat surface and then formed the base for new deposition may not be so easy to discern. In this case one has to look at the rock type, are the environments of deposition greatly different? or if there is a thin layer of weathering material between the two periods of deposition.

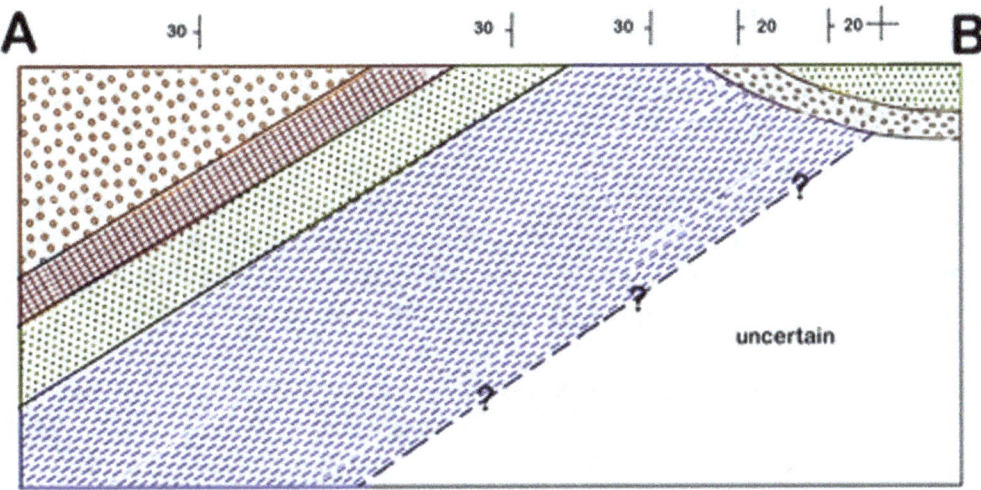

CONCLUSIONS:

1. **What types of unconformities are shown in the cross-section (the unconformity symbol is not shown here)?**
   The basalt came from a lava flow on the surface on top of the sandstone. After a period of erosion (?) the conglomerate was deposited. This is a paraconformity as the basalt layer and the sedimentary beds above and below are parallel. Closer to B is a typical angular unconformity where the shale has been eroded to a basin around

B and then filled in with new sediment – typically a coarse conglomerate of eroded material from the basement, followed by sand as the water flow decreased.

2. What is different about the basalt? How did it get there?

It is tilted like the sedimentary beds above and below it. This means that it formed as a horizontal lava flow on top of the horizontal sandstone and then covered with the conglomerate. At a later stage, all beds were then tilted at 30^0 towards A.

3. Which beds were deposited (a) first and (b) last?

The shale bed closest to B was deposited first as muds, and then came the other sedimentary rocks in sequence, the basalt and then the conglomerate. All were then tilted towards A but near B the shale was eroded out as a basin and then the conglomerate and then lastly the coarse sandstone were deposited.

RESEARCH (Optional):

Unconformities represent a break in the geological process, often with uplift and erosion. They herald a new phase in the geological history. Some of the main types are given in:

http://www.indiana.edu/~geol105b/images/gaia_chapter_6/unconformities.htm

http://homepage.smc.edu/grippo_alessandro/unc.html

Experiment 3.7: Geological History

One of the main reasons for mapping an area is to work out its geological history, the sequence of events which led to its geology. This not only includes the sequence of deposition of sedimentary strata but also any igneous intrusions, subsequent metamorphism as well as evidence for uplift and erosion. In reality, field geologists will make use of whatever data is available including drill cores from work being done in the field or from libraries of past drill cores. They may also enlist the aid of geophysics for any subsurface data and make use of satellite and aerial surveys.

This exercise contains several elements studied in the past: inclined beds with apparent dips, a fault and a massive intrusion. Calculate the apparent dip of the layered sequence which dips at 55^0 towards a direction of 50^0 (direction angle) from the A-B traverse. Consulting the tables in Experiment 16.1, the apparent dip of 43^0 is found. The fault is drawn in first with its dip of 70^0 towards A making sure that when the bedding planes are drawn in, that they do not cross the fault. One can also note that the massive granite (a coarse-grained igneous rock typical of some batholiths) intrudes through the fault suggesting that the faulting occurred first. Draw in the batholith as a massive region getting bigger with depth. Notice that this map does not show a small aureole around the batholith as it is so massive that all of the original sedimentary strata have been metamorphosed to the rocks shown on the map. Additional construction is also needed to obtain the positions of some of the beds to the left of the fault.

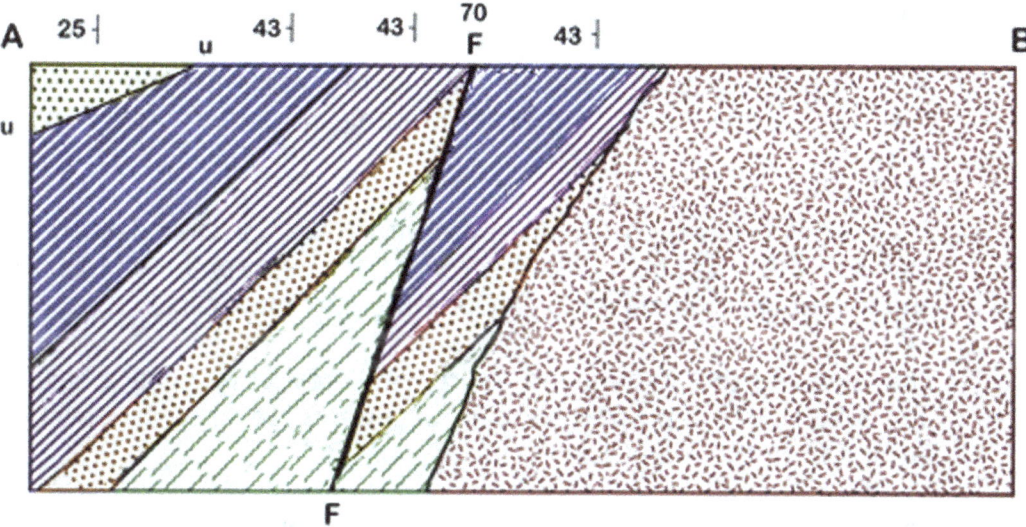

CONCLUSIONS:

1. **What types of unconformity and fault are shown in the cross-section?**
   This cross-section shows a nonconformity and a normal fault.

2. **What is the (a) heave and (b) throw of the fault?**
   Taking the bottom of the phyllite as a marker bed, the heave and through are each about 500 m.

3. **What is different about the layered beds? How were they originally formed?**
   They have been metamorphosed by the massive granite intrusion. Marble was once limestone, the calcium-silicate hornfels was a calcium rich mudstone and the hornfels was a mudstone or shale. The quartzite was once sandstone but the schist is a regional metamorphic rock whereas all of the others were formed from contact metamorphism. It was probably there before the other rocks were deposited and may represent an early sequence of rocks eroded and now forming a paraconformity.

4. **LIST in order all of the events which occurred in this region from the first event to the current day (CARE: this is a complex problem so requires some research and in-depth description).**
   The geological history was probably:

   a) The schist was originally a shale or mudstone which was regionally metamorphosed by extreme heat and pressure from an earlier tectonic event.

   b) This was uplifted, by tectonism, and eroded to a flat plain.

   c) After an hiatus, new sedimentation, probably following subsidence and a marine incursion, sand was deposited. This was followed by muds, then calcium-rich sediments and then limestone. It is probable that the limestone represents an old coral reef.

d) More tectonism produced the tilting of these marine sedimentary beds towards the northwest and later faulting occurred as a result of tensional forces.

e) A large intrusion by a granite batholith metamorphosed all of the rocks by contact with its heat. The strongly metamorphic schist probably only underwent slight changes to produce the current rocks.

f) There was further tilting and uplift with erosion forming a basin in the west into which was deposited sand in a new period of sedimentation, probably eroded from the granite.

g) The entire region was again uplifted and completely eroded to a flat surface.

This last step is often forgotten by students.

Experiment 3.8: Half-life of Radioactive Iodine 131

This is a paper activity to show the features of half-life and how it can be used in determining the age of minerals, rocks, gases and other materials by the various isotopes contained within them. Iodine-131 is a radioisotope of iodine and has a radioactive decay half-life of about eight days. It is a radioactive isotope present in nuclear fission products of uranium and plutonium and was a significant contributor to the health hazards from open-air atomic bomb testing in the 1950s, from the Chernobyl disaster and from the Fukushima nuclear crisis.

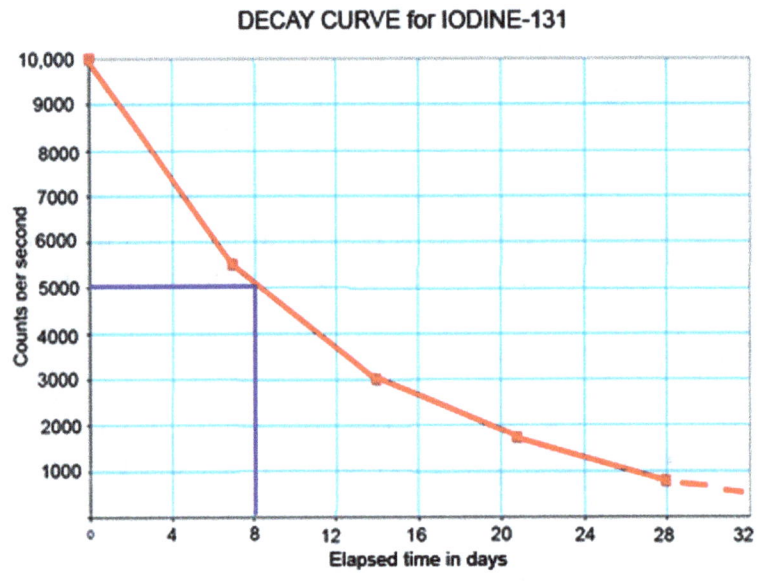

CONCLUSIONS:

1. **What is the graphical value of the half-life for iodine-131?**
   8 days.

2. **What could be some of the errors involved in this experiment?**
   For this activity the main error is in designing and plotting on the graph. In real life there would be an error in the natural background radiation, the ability of the radiation counter to refresh after every count and the error in reading the digital counter.

3. **If 40 grams of iodine-131 was used at the start of the experiment, how long would it take for this mass to decay to 5 grams of iodine-131?**
   With a half-life of 8 days, there would be 20 g after 8 days, 10 g after 16 days, and 5 g after 24 days.

RESEARCH: (Optional)

Iodine-131 is used in the medical industry especially in the treatment of cancers of the thyroid gland or a hyperactive thyroid (thyrotoxicosis) in the neck because this gland absorbs iodine. The radioactivity in controlled doses destroys the cancer.

Experiment 3.9: Size of the Earth

This is a neat version of the method used in 240 B.C. by the Greek astronomer Eratosthenes of Cyrene (now modern Aswan, Egypt) by noting the angles of shadows in Alexandria and Cyrene further south down the Nile on the summer solstice. It is a good experiment to show how the ancient world had some clever ideas and did some remarkable science with the best mathematics and equipment at the time – and a lot of good, practical observation and common sense!

In this experiment, the length of the shadow at noon (or use solar noon and the length estimated at that time by plotting length and time as in the previous experiment) can be measured and its angle calculated using:

Tangent of the angle = actual length of the shadow stick/length of shadow

finding the angle for this tangent using a tangent calculator or the web site at: https://www.rapidtables.com/calc/math/Tan_Calculator.html which has a good inverse tangent function (tan. to angle)

Then, instead of travelling to a place north or south of one's location, find a location on a map (or Google Maps) which is on the equator exactly due north (for the Southern

Hemisphere) or due south (for the Northern Hemisphere) of one's location and then go to the solar calculator website at:

https://www.suncalc.org/#/-4.2,152.1645,15/2018.04.27/14:00/1/0

type in the location on the Equator to get the angle of the Sun at the same time.

Then use an atlas (or Google) to measure the difference in distance between the Equatorial location and the home location. This distance will (like that between Alexandria and Cyrene) represent that part of the Earth's curve representing the difference in Sun angle between the two places. This can be then scaled up to get a value for the circumference of the Earth.

For example: Brisbane, Australia, the length of shadow for a 61 cm shadow stick was 53.5 cm. This gave a tangent of 61.0/53.5 or 1.1401. Using the calculator this tangent is for an angle of 48.7 degrees.

Now at the same time, the angle (using the Solar calculator site) at Rabaul (Papua New Guinea – at 4^0 North, the closest centre of population near the Equator due north of Brisbane) was 71.8^0 and its distance from Brisbane is 2587 kilometres.

i.e. an angular difference of 23.1^0 representing a change in distance of 2587 km. Therefore, for a change in angle of 360^0 the distance (circumference) should be:

$$\frac{2587 \times 360}{23.1} = 40,316 \text{km}.$$

The actual polar circumference of the Earth is 39,931 km i.e. an error of + 385 km or 9%

This was due to the errors in measuring the shadow and shadow stick (+/- 0.5 cm) and the slight digression from Rabaul as being due north (it is one degree of Longitude and 4 of Latitude out).

See tangent converter at (use the reciprocal to get angle)
https://www.rapidtables.com/calc/math/Tan_Calculator.html

and the Solar Angle calculator for many locations is at:

https://www.suncalc.org/#/-27.4698,153.0251,14/2018.04.27/12:00/1/0

CONCLUSIONS:

1. What is the calculated value for the circumference of the Earth (along North South)?
   As per student calculations but also see above.

2. How does this compare with the actual value of 39,931 kilometres? Calculate the percentage error for this experiment.
   See students' calculations and also the above calculations.

3. How could this experiment be improved?
   Use a longer shadow stick and better measurements (+/- 1.0 mm) and solar noon not 12 noon.

3.4 Other Activities

- Obtain a geological map and sequence the geological history of the local district. Find out how the rocks and ages have been correlated with type areas in Europe.

- Invite a local geologist to give a talk on how modern geological maps are drawn

- Obtain information about local geology and mapping from the USGS at:

   http://www.earthsciweek.org/geologic-map-day/learning

- If possible, go to a nearby outcrop, do a simple linear open traverse and map the rocks.

- Make a model in a large, flat box using different coloured clay in layers and as intrusions and then plane off the top to give a simulated geological area which the students can explore along different traverses of their selection. Drill cores may be taken using small diameter pipes and the subsurface geology is proposed. If the box has at least one side made from glass (use a small fish tank) or acrylic then the actual subsurface geology can be shown.

3.5 Answers to Multichoice Questions

Q1.B Q2.D Q3.C Q4.B Q5.C Q6.D Q7.A Q8.A Q9.B Q10.C

3.6 Some Suggestions for the Review Questions

1. State the laws or principles of:

   a. **Original Horizontality** - layers of sediment were originally deposited horizontally under the action of gravity.

b. **Lateral Continuity** - layers of sediment initially extend laterally in all directions.

c. **Superposition** - in any undisturbed sequence of rocks deposited in layers, the youngest layer is on top and the oldest on bottom.

d. **Faunal Succession** - fossils succeed each other vertically in a specific, reliable order that can be identified over wide horizontal distances.

2. **In this modern age of electronic information devices, why is it still important to understand the basics of geological mapping on paper? Discuss.**
For small scale projects it may be necessary to map the area in the traditional manner. Students should always understand what the modern technology is attempting to do and some research students may not have modern facilities and have to map in the traditional manner.

3. **Some geologists prefer to use dip direction instead of strike in their field measurement of tilted beds. Discuss the pros and cons of this preference.**
This is usually the most convenient way, especially if only a small exposure of bedding plain is available. It is a simple process to place the compass horizontally above the dipping plane, measure the dip direction then quickly turn the compass on its side still facing that direction and measure the angle of dip.

4. **Look at the photographs of the Brunton pocket transit in the chapter in the textbook. What are the instrument errors of the:**

   (a) clinometer scale and
   (b) azimuth scale (for geographical bearings)

   Both are +/- 0.5 of a degree

What other errors could occur in the field in using the Brunton pocket transit?
The most common one is dirt or mud on the face. Also, rain, snow etc. can make readings difficult when readings are taken in bad conditions. Parallax error occurs if the geologist is looking at the scales at an angle and sunlight reflecting off the face is also a problem.

5. **A traverse along flat ground encounters dipping strata which is not dipping in the same direction as the traverse. The beds are all dipping at 30^0 (i.e. apparent dip). What is the true dip if the:**

   (a) angle between the dip direction of the beds and the traverse
   is 20^0? Refer to the table in Experiment 16.1. This answer is about 32^0

   (b) angle between the strike of the beds and the traverse is 60^0? About 49^0

6. **On another traverse, a bed of limestone dipping (true) at 40^0 outcrops for 120 metres along the traverse. What is its true thickness? How could that be confirmed?**
Some basic trigonometry needed here or do a drawing to scale but remember that thickness is measured at right angles to both bedding planes. The answer is approx.

80 metres. This can be confirmed with looking for a natural cross-section of the bed in a cliff face or selective drill coring.

7. This question refers to the following graph made by measuring radioactive decay counts (as clicks on a radiation detector) against time for a particular isotope (type of atom of an element).

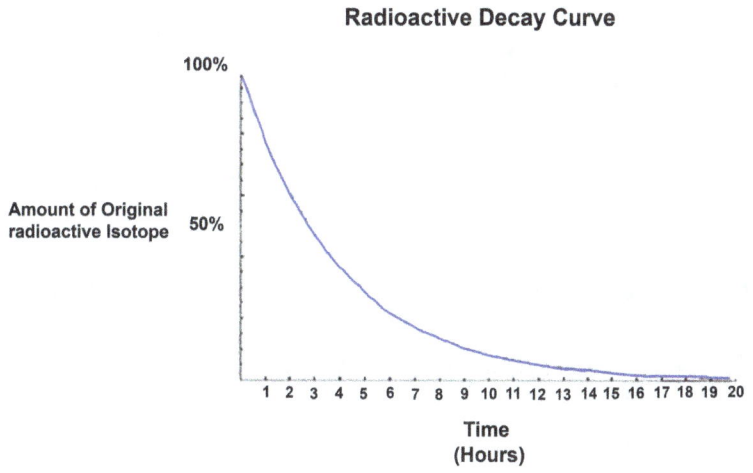

What is the Half-life of this radioactive isotope?
Just under three hours.

8. What are the probable past environments indicated by the following rocks:

   (a) **greywacke** - a deposited formed from compaction of a turbidity current off a coastal marine platform.

   (b) **volcanic breccia** - from an explosive volcano nearby (so that only the big clasts fall).

   (c) **tillite** - from moraine from the sides or base of a glacier. Glacial varve shales come from finer sediments deposited from the meltwater streams of glaciers.

   (d) **mudstone with trilobite burrows** - trilobites were bottom dwellers (mostly) in shallow seas. The mudstone has come from muds in the shallow but quieter parts of oceans.

9. This question refers to the following geological cross-section:

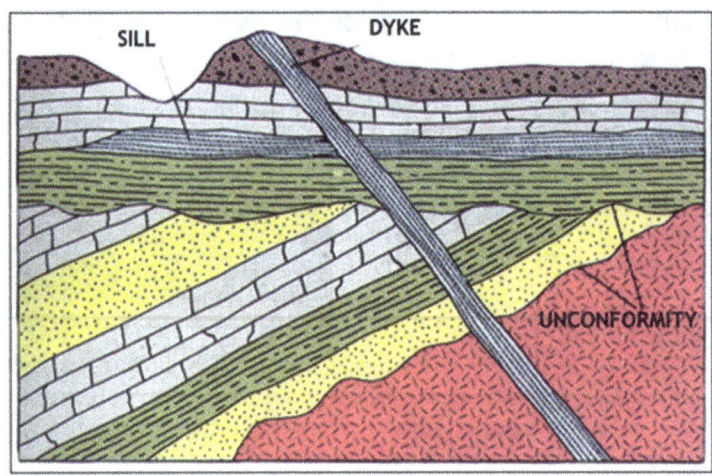

Give a sequential geological history of this area from oldest to youngest.

Knowledge of the basic rock symbols is useful here. A probable sequence could be:

- the intrusive igneous rock, bottom right, had been uplifted and eroded, paraconformity.

- this was covered was covered by a sedimentary sequence, probably marine because of the beds of limestone, of sands, muds, coral reef deposits which became in turn sandstone, shale, or mudstone, limestone, more sandstone, if the land sank and a marine transgression allowed for more sand to be deposited, and then more limestone, more reef activity.

- This was all tilted to the left, uplifted and eroded to a flat surface, angular unconformity.

- A new marine sequence was deposited as shale or mudstone followed by limestone and a coarse conglomerate, or sedimentary breccia from a nearby coastline.

- then the upper sedimentary sequence was intruded by a sill which moved between the bedding plane of the weaker mudstone and the limestone above.

- The entire sequence was then intruded at a later stage by a dyke.

- The surface was then eroded to the present topography.

10. **Explain how the contributions of the sciences of seismology and astronomy helped to establish a model for the Earth's interior.**

Isaac Newton calculated, from his studies of planets and the force of gravity, that the average density of the Earth is twice that of surface rocks and therefore that the Earth's interior must be composed of much denser material. One hundred years later, Emil

Wiechert, realized that the density of the Earth is greater than the density of rock, meaning that the Earth cannot be composed entirely of rock. He suggested that the Earth might be like a giant meteorite with a core of nickel-iron metal which had settled to the centre. In 1909, the Croatian seismologist Andrija Mohorovičić, observed that seismograms from shallow-focus earthquakes had two sets of P-waves and S-waves, one that followed a direct path near the Earth's surface and the other refracted by a high-velocity medium which represented the denser material below the less dense material of the direct-path waves. This was the boundary between the Crust and the Mantle below. Beno Gutenberg, found that P waves travelled through the entire body of the Earth but S waves did not. He found that the S waves were stopped at a boundary 2,900 km within the Earth and as S did not travel through liquids, he concluded that the outer part of this core was liquid. Later, the Earth's inner core was discovered by Inge Lehmann who noticed that P waves travelled directly through the central part of the core of the Earth which must be a solid.

3.7 Reading List

Iowa State University (2015). *Glossary of Geologic Terms*. Dept. of Geological and Atmospheric Science. http://www.ge-at.iastate.edu/glossary-of-geologic-terms/

Janke, P.R. (1996). Correlating Earth's History - Lecture Notes. From a lecture given in May, 1996: *Learning from the Past*. Black Hills Museum of Natural History. http://www.wmnh.com/wmas0002.htm

Kent Educational, (2015). *Careers in Geology*.
http://www.personal.kent.edu/~cschweit/Stark/whygeologymajor.htm

Lisle., R. J., Brabham, P. & Barnes, J. (2011). *Basic Geological Mapping* (5th Edition). Oxford. John Wiley & Sons. 235 pp. ISBN: 978-0-470-68634-8

Lisle, R. J. (200). *Geological Structures and Maps: A Practical Guide*. (3rd. Edition) Oxford, UK. Elsevier Butterworth-Heinemann. 106 pp. ISBN 0 7506 5780 4

Mathez, A. (2000). *Earth: Inside and Out*. New York: New Press, American Museum of Natural History. Excerpts at:
http://www.amnh.org/education/resources/rfl/web/essaybooks/earth/

Roberts, J. L. 2013. *Introduction to Geological Maps and Structures*. Pergamon International Library of Science, Technology, Engineering and Social Studies.332 pp. ISBN: 9781483140995.

Soller, D. R. (2004). *Introduction to Geologic Mapping*. USGS. (first published in the McGraw-Hill Yearbook of Science & Technology 2004, pp. 128-130.).
http://ncgmp.usgs.gov/geomaps/introgeo_mapping.html

Tearpock, D. J. & Bischke, R. E. (2002). *Applied Subsurface Geological Mapping with Structural Methods*. Upper Saddle River, NY. Prentice-Hall. 864 pp. ISBN: 0-13-859315-9.

Additional References

https://serc.carleton.edu/NAGTWorkshops/structure/resources.html   Great links to modern mapping references.

http://www.virtualmuseum.ca/edu/ViewLoitLo.do;jsessionid=F59A7F1E5CA89D15D9A60898058D109C?method=preview&lang=EN&id=26053   many links to the New Brunswick Museum site

https://www.e-education.psu.edu/earth106/content/l4.html good links from Penn State University

Chapter 4: Rock-forming Minerals

4.1 Theme

The Geosphere consists of rocks which are themselves made up of mostly two or more minerals. Most of the minerals of the Earth are silicates consisting of the elements silicon, oxygen and other elements such as sodium, potassium, iron and magnesium.

4.2 Rationale

Minerals are the basic building blocks of rocks and many minerals are valuable in their own right. There are many careers associated with minerals such as Mineralogists, Geochemists and Exploration Geologist who look for economic minerals. They are also interesting and good crystals which are hard to find that can be very valuable for the amateur collector.

4.3 Notes on the Practical Work (see PRACTICAL MANUAL 1.)

Experiment 4.1: Some Common Rock-Forming Minerals

A good introduction for this practical activity is to show students a variety of specimens containing visible crystals of common minerals e.g. large specimens of intrusive igneous rocks such as pegmatite, granite and gabbro. Other instances of crystals such as uncut or cut gemstones and the metal crystals seen on galvanized metal are also useful.

Whilst hand lenses are a basic way of closely examining specimen and their use should be practiced at every opportunity, a stereo microscope, also called a dissecting microscope by Biologists, or a binocular microscope, is very useful for class observation.

These are usually about 20 power (x20) and are easy to use. They also give an excellent three-dimensional view of any specimen. Cheap versions can be had for under $100 and a class set of (say) ten can be kept in the laboratory for group use. The main issue with these microscopes is that students must be taught to place the specimen on the stage carefully and in a well-lit area. Microscope lamps can also be used, and then to focus upwards when looking through the eyepieces. Focusing downwards usually results in grinding the objective lens into the specimen.

For this experiment, a class set of common rock-forming minerals is needed. These could be individually labelled (e.g. M1. M2 etc.) and boxed in small, labelled (Minerals 1, Minerals 2 etc.) containers. Mineral specimens should be small, hand specimens about 4-5 cm across and clearly show well-formed crystals or good cleavage (as appropriate) and free of weathering. Basic minerals include: quartz, orthoclase feldspar (pink), a plagioclase feldspar (white), biotite mica, muscovite mica, hornblende, augite, olivine (usually as granules), calcite and barite. This list may be supplemented or substituted but should not be too exhaustive for the students to comprehend in one study. About ten is satisfactory. Notice that hornblende and augite and calcite and barite are included as these are superficially similar and often confused. Students should see that testing for a wide range of properties leads to a more accurate identification.

Also required for this activity are streak plates and Mohs' scale kits. Rather than purchasing overpriced kits from suppliers, teachers can get cast-off small, white (and black) bathroom tiles (say about 5 cm square) and use the under surface (unglazed) for testing. Similarly, a class set of Mohs' Scale kits can be made up using smaller mineral fragments numbered 2-9 (talc and diamond are probably not useful here) e.g. gypsum (2), calcite (3), fluorite (4), apatite (5), a feldspar (6), common vein quartz (7), topaz (8) and common corundum (9). The latter two minerals may not be readily available and in most cases not needed, quartz being hard enough, any mineral which will scratch it will be of some interest (gem hardness). A small plastic container is very useful for these kits but they are often mixed up with the mineral specimens so labelling should be distinctively different. A liquid paper or white paint circle with numbers painted in red is most useful.

HARDNESS is a good test and so students are encouraged to do it right. They should find a good flat surface (probably a cleavage plane) and then drag the Mohs' mineral across it with some downward pressure to try to make a scratch. If a line of powder is seen, then it should be rubbed away with a finger to see if there is a scratch below or if it is simply dust from the Mohs' mineral. Conversely, if no scratch is made (this can usually be felt), then the student should attempt to scratch the Mohs mineral with the specimen. A good exercise (especially in a practical exam) is to give students a set of white minerals (e.g. white plagioclase, calcite, gypsum, vein quartz, and kaolinite) and tell them to find a way to identify them. They can do this by scratching each mineral with another until an order of hardness is obtained (gypsum, kaolinite, calcite, plagioclase, and quartz).

For **CLEAVAGE**, students are encouraged to estimate this by <u>inspection</u> with a hand lens (good training for field work) or with the stereo microscope. The cleavage of these specimens using a hammer is <u>not</u> encouraged. Also, students should be trained in the philosophy of honest observation. Even if a mineral is known to have a property but it cannot be seen in the specimen, the true observation must be recorded. In doing this activity, results for several tests on a specimen should be enough to give an accurate identification even if one or two of the observations is inaccurate or unable to be made.

SPECIFIC GRAVITY testing requires more detailed use of balances and other equipment. This is best left to another, separate experiment. Heft can be generally used here as light, medium or heavy. Most specimens here will fall into the medium category although the micas will be light and the barite easily distinguished from the calcite by being heavy.

A **SKETCH** is most important and the suggestions for sketching are given in the practical notes. It is not an art class, so students should be encouraged to draw a representation by looking at the most important features and then exaggerating them in their outline of the specimen. Sketches should be about one-half page across, fully labelled and with some scale. If possible, coloured minerals should be represented by a true representation in the sketch. A small, cheap set of coloured pencils is recommended. Students using electronic note-taking devices (tablets, iPads etc.) are encouraged to make pencil/paper sketches in class and then scan them at home; few students have the artistic and technical skills to use computer art programs quickly and accurately (but don't discourage those who are good and fast). Some students may wish to use their cell phones or tablets instead of drawing the specimens. This is good for more elaborate situations (such as field structures etc.) but often the depth of field, lighting and camera angle are not good. Learning the skill of doing a quick, accurate and colourful scaled diagram is a big advantage.

CONCLUSIONS:

1. **LIST the code numbers of the specimen, the specimen's names and at least two DISTINCTIVE properties of each mineral which will then help in the quick identification of that mineral.**

   As per the activity. After the experiment has been written up (and marked if appropriate), the teacher should correct any mistakes so that students will have some KEY WORDS to remember for each.

2. **COMMENT on the use of mineral properties (advantages/disadvantages) in identifying unknown minerals.**

   One of the main skills of the geologist! In the field, these skills for both rock-forming minerals and ores are most important. Advantages includes obtaining a quick identification and therefore estimation of the mineral deposit, ore or rock but the disadvantage is that it is usually difficult to find good, clear mineral surfaces in the field upon which to do the tests. Usually a hand lens will give some idea of colour, cleavage and hardness (using a field scale or just a piece of common quartz).

3. **What are some likely errors which may limit correct identification of minerals in the field?**

   Weathering and generally limited exposure will create problems, but the good field geologist also has an idea of the weathering products of many of these minerals.

Experiment 4.2: Growing Crystals

This is both a useful and fun experiment about crystals.

PROCEDURE A: uses solid salicylic acid which dissolves readily in hot water but crystallizes in cold water.

It has the chemical formula of HOC_6H_4COOH and its name is derived from the Latin _salix_ for _willow tree_ as it was once extracted from the bark of these trees (by the ancient Greeks, Romans and Druids and other ancient herbalists) for curing headaches and other pains. It did, however give some stomach problems so later, chemists made it more useable by making its acetate (acetylsalicylic acid) which we call _aspirin_.

CARE! The main danger from this experiment is from scalding from using hot water or burns from carelessness from using the burner. Students should be warned about bumping of the hot water if heated too fast and NEVER point the test-tube towards others. A demonstration of how to heat small amounts of water in a test-tube by using a low blue flame and heating the TOP of the water would be useful first. Using salicylic acid in this experiment is quite safe if not ingested nor rubbed onto the skin. It will cause nausea if ingested and high concentrations will irritate and peel skin (it is used to remove warts).

QUESTIONS from the Experiment:

1. **What happens as the solution cools? Why?**
   Needle-like crystals of salicylic acid form as they are less soluble in cold water. This occurs in nature when crystals form from hot, volcanic solutions. Solubility of salicylic acid is:

   $$1.24 g/L (0°C)$$
   $$2.48 g/L (25°C)$$
   $$4.14 g/L (40°C)$$
   $$17.41 g/L (75°C)$$
   $$77.79 g/L (100°C)$$

2. **Describe any shapes. Sketch the apparatus and observations in the Results section.**
   The salicylic acid will suddenly appear as long, transparent crystals which can be drawn as they are seen within the test-tube.

PROCEDURE B: Safety note – this part uses copper sulfate crystals and its solution. Both are poisonous and CARE is needed so that they are not ingested nor make contact with the skin (which will die and peel off). With gloves and proper care this experiment is quite safe. Instead of using $CuSO_4$, alum (or chrome alum) can be used. These are safer but often more difficult to obtain good crystals as they tend to absorb moisture from the air.

The initial solution needs only to be strong enough in concentration to give good crystals (overnight) by evaporation. After that, any solution which is used to grow the bigger crystals must be <u>saturated</u>, i.e. the water solvent must contain all of the solute crystal that it can at that temperature.

The first part of this experiment (obtaining seed crystals) is best done at the end of Procedure 1 for use the next day. Students can put their petri dishes into a covered area such as a fume hood if available but they must be exposed to air and out of the way. Watch glasses can be used but the petri dishes give better uniform depth and so a better chance of good crystals.

The second phase requires students to select a good crystal (best <u>shape</u> not size) and lasso it with fine thread. Thin copper wire, similar to that used in winding electric motors, is the best to use (see the Physics Faculty or electrical components store). CARE with the handling of these crystals. Use disposable surgical gloves if they are available. Tie the thread onto a stick or glass rod so that it will hang into a small 150 ml beaker as shown.

In the next phase it is important to use a saturated solution of the crystal, otherwise the seed crystal will dissolve in the excess water of the solution. It is a good idea for the teacher to make such a solution by dissolving a good amount of crystal in hot water (until no more dissolves) and then storing it in a large bottle. The old brown Winchester Bottles of about four to five litres are excellent. Once cooled, the solution should be saturated and there will probably be large crystals on the bottom of the bottle. Enough of this is needed for the whole class (say one beaker/group) over several days. Every day, have the students top up their beakers so that their crystals remain below the solution. Results should occur overnight.

CONCLUSIONS:

1. **Comment generally on how these crystals can grow from solution and how smaller crystals grow to bigger size.**
   They grow by accretion as the crystal edges grow from addition of the ions from the solution building up the crystal lattice. Teachers may award a prize for the best shaped crystal.

2. **What are the limitations to the size of the crystals?**
   Usually only the size of the container and the consistency of the saturated solution. With care, very large crystals of several centimetres across can be grown. Once they become too big for the beaker, they can be transferred to a broad and deep container such as a bowl. This could be an ongoing class activity over several weeks.

3. **Is there any indication within the crystal as to how they grow from smaller crystals?**
   Careful examination with a hand lens or stereo microscope will show concentric growth from the seed crystal outwards. There may also be some aberrations of extra crystal growths growing or twinning out from the original. This should show how the crystal grows outwards by accretion.

4. **Where in Nature do crystals grow in this way (Internet research needed).**
   This is crystal growth by cooling from a saturated solution. There are many natural examples such as those crystals growing in hot springs and from superheated water in veins at the end of magmatic emplacement into country rock. Do not confuse this from the growth of crystals from molten magmas nor from crystals formed by evaporation (but this does relate to the growth of the seed crystal).

Other Activities possible in support of this experiment:

1. **Demonstrate the crystallization of the chemical Salol** (phenyl salicylate - M.P. of 41.5°C /106.7°F) by melting a small amount in a disposable pyrex glass or metal dish and then allowing it to cool at room temperature to show how crystals can form from molten material.

   Alternatively, this can be demonstrated nicely if the classroom has a microscope with a video camera attached to a computer or data projector. Place a drop of the molten Salol onto a warm, clean microscope slide and place it in the field of view. Watch the crystals forming and note their aggregation and shape.

2. **Demonstrate the effect of rate of cooling** from the molten state using sulfur powder. In a clean test-tube gently melt the sulfur (at 114°C). Pour some of it into

a glass beaker of cold water. Allow the remainder in the test-tube to cool at room temperature. Compare the crystalline nature of the sulfur in the test-tube with the hard, non-crystalline sulfur which has formed as plastic beads (they bounce!) in the water. The sulfur in the test-tube is monoclinic sulfur whereas that in water is amorphous sulfur. The powder taken from the supply bottle is the more stable form of orthorhombic sulfur and these three forms are the main allotropes of the element.

CARE: Do not heat the sulfur too much nor place the powder near the flame as it will ignite to give choking sulfur dioxide gas.

3. **Grow Metal Silicate Crystals** in an <u>old</u> glass jar with a good sealing cap (not a good beaker etc.) add some pure quartz sand as a small base then carefully fill with a solution made from a 1:4 mixture of Waterglass (sodium silicate) to distilled water. Add some good-sized crystals of any of the following chemicals:

    > cobalt chloride (gives pink crystals)
    > manganese chloride (purple)
    > iron iii chloride (orange)
    > nickel nitrate (green)
    > copper sulfate (blue) and
    > calcium chloride (white)

    These react with the silicate to give the metal salts of this compound and students can watch this happening in real time as the crystals grow as long strands upwards. This crystal garden will not last long as the solution goes cloudy with exposure to the air. It can be made to last for years if water is SLOWLY trickled into one side to flush out the excess silica solution. Place a good sealed cap onto the top and leave where it will not be disturbed.

4. **Construct lattice models** of some common materials e.g. sodium chloride and some carbon rings (as in graphite). Show professionally constructed models of diamond, graphite and any other compound (see the chemistry faculty).

5. **Student construction of cardboard crystals** cut out from sheets and then folded and taped.
    See:

    http://www.ellenjmchenry.com/homeschool-freedownloads/earthscience-games/documents/Crystalshapes.pdf

    https://www.thoughtco.com/make-your-own-magic-rocks-607653

    www.earthsciencewa.com.au/mod/resource/view.php?id=1600

Experiment 4.3: Measuring Specific Gravity

Specific Gravity (S.G.) is a good laboratory diagnostic for minerals. In the field, the heft (or general idea of relative weight/size of the specimen) is sometimes useful, especially if the mineral has a high S.G. such as baryte.

This method uses **Archimedes' principle** which states that the upward buoyant force that is exerted on a body immersed in a fluid, whether fully or partially submerged, is equal to the weight of the fluid that the body displaces and acts in the upward direction at the centre of mass of the displaced fluid.

Archimedes of Syracuse (c. 287 – c. 212 BC) was a Greek mathematician, physicist, engineer, inventor, and astronomer who lived in this Greek city in what is now Sicily, Italy. The story goes (according to Vitruvius, a Roman architect and engineer) that a gold crown was made for King Hiero II of Syracuse as an offering for a temple. The king, who had supplied the pure gold to be used, asked Archimedes to determine whether some cheaper silver had been substituted by the dishonest goldsmith. Archimedes had to solve the problem without damaging the crown, so he could not melt it down into a regularly shaped body in order to calculate its volume and therefore its density. While taking a bath, he noticed that the level of the water in the tub rose as he got in, and realized that this effect could be used to determine the volume of the crown. For practical purposes water is incompressible, so the submerged crown would displace an amount of water equal to its own volume. By dividing the mass of the crown by the volume of water displaced, the density of the crown could be obtained. This density would be lower than that of gold if cheaper and less dense metals had been added. It was rumoured that the dishonest goldsmith was punished.

CONCLUSIONS:

1. **What was the calculated value for S.G. and its error?**
   Calculate the separate instrument errors (one half unit for analogue scales and one unit for digital scales) as a percentage of the value measured and then add these errors.

2. **From a list of S.G. what is this mineral?**
   As appropriate to the value found and specimen used +/- % error
   Some common specific gravities are:

Graphite	C	2.23
Quartz	SiO_2	2.65
Feldspars	$(K, Na)AlSi_3O_8$	2.6 - 2.75
Fluorite	CaF_2	3.18
Topaz	$Al_2SiO_4(F,OH)_2$	3.53
Corundum	Al_2O_3	4.02
Baryte	$BaSO_4$	4.45
Pyrite	FeS_2	5.02

Galena	PbS	7.5
Cinnabar	HgS	8.1
Copper	Cu	8.9
Silver	Ag	10.5

3. **How does the experimental value agree with the book value for this mineral?**
   Depends on the ability of the student and the purity of the mineral. Baryte is better than galena as the latter often is mixed with sphalerite which has a lower S.G.

4. **Comment on any student, environmental or instrument errors.**
   Discuss instrument error and purity of specimen as well as carelessness in observation and measurement.

5. **How could this experiment be improved?**
   Use a bigger specimen, a better balance and ensure purity of specimen.

Experiment 4.4: Geochemistry

This can be a complicated experiment and with some larger, difficult classes the teacher may wish to do part or all of it as demonstrations. With good classes it is still complicated but a worthwhile experiment, linking Geology with Chemistry and showing the usefulness of chemical analysis in mineral identification.

CARE! Assume that all of the chemicals and specimen to be used are toxic so warn the students about ingesting material and handling chemicals. Use of the burners with a naked flame can be dangerous so students should wear non-flammable aprons if possible.

PROCEDURE A: Flame Tests. This is usually a lot of fun for the students and relates to their knowledge of fireworks. Ordinary paper clips can be bent so that they can be held at one (looped) end and the other (straight) end can be dipped in water and then into a SMALL container (e.g. small cylindrical pill bottle) of the metal salt. Care must be taken to thoroughly wash the end of the wire in water to prevent contamination.

The edge of the BLUE FLAME of the burner is used as it is relatively colourless and very hot. The colour of the first flame is to be noted. Some of the colours should be:

COPPER	rich green
SODIUM	yellow
BARIUM	pale green
STRONTIUM	crimson
LITHIUM	lilac

Other Activities in support of this experiment:

1. **Other colours** can be obtained from calcium salts (orange), potassium chloride (purple), copper chloride (blue), magnesium sulfate (white) and powdered iron (filings) makes many sparks (fun).

2. **Try mixing a couple of salts together** and get the students to identify the components. Later in the Astronomy chapter this will be useful in explaining how elements are detected in stars using spectroscopes.

3. **Show some discharge tubes** (see the Physics Faculty) of some of the elements e.g. hydrogen, helium, mercury.

PROCEDURE B: Chemical Tests. These are more complicated and require good self-control by the class members and vigilance by the teacher.

CARE! Acids are corrosive and will harm eyes and sensitive skin even when diluted. 2 Molar hydrochloric acid is used here (as sulfuric acid may complicate issues with sulfides and sulfates). But the teacher may wish to experiment with weaker dilutions. If necessary, some or all of this experiment may be demonstrated and the test-tubes taken around the class for students to smell. Note that the test for sulfides will give toxic hydrogen sulfide gas which has the smell of rotten eggs. If the acid strength used is too dilute, the reaction will not occur. If so, the teacher can demonstrate this reaction by cautiously warming the reaction mixture and then taking the test-tube around the class for a brief time. There seems to be a change in reaction rate when this mix is washed out in water (increases) so have a fume hood handy or a bucket nearly full of water in which the test-tubes can be placed and then carried outside. Good ventilation is needed in the room as even a small amount of H_2S will make some students nauseous.

Also, take care with silver nitrate solution also reacts with chlorides in the skin producing black spots which cannot be erased or washed off. One has to wait until that layer of skin naturally dies and peels off! It will also react with some paints (on walls and desks) and material dyes to give permanent black spots. The students should be cautioned that silver nitrate is toxic so that only one drop should be applied to the chloride solution using a good dropper from a dropper bottle clearly labelled. The reaction with the chloride solution will be a white cloud of silver chloride. This is photosensitive (the silver nitrate and its solution should be in brown bottles) and the teacher can demonstrate this reaction using a beaker of chloride solution and a few drops (say 10) of silver nitrate and then filtering it through filter paper in a filter funnel. This can be placed in sunlight with some object (e.g. an old key) on top. Within a short time, the exposed silver chloride will turn purple (white Chloride + black free silver) and leave a white impression of the key.

The appropriate reactions of each of the tests are:

CARBONATES – odourless, colourless carbon dioxide gas is given off as bubbles. The teacher may demonstrate that this is really CO_2 by bubbling some into limewater which turns milky e.g. with calcite:

Calcium Carbonate + Acid = Carbon Dioxide gas + Calcium salt of that acid

CHLORIDES - white precipitate of silver chloride which may turn purple due to black free silver forming with the influence of light breaks up the silver chloride e.g.

Sodium Chloride + Silver Nitrate = Silver Chloride + Sodium ions and Nitrate ions.

Silver Chloride + light = Silver + Chloride ions

SULFATES - a dense white precipitate of Barium Sulfate forms with the Barium Nitrate solution e.g.

Magnesium Sulfate + Barium Nitrate = Barium Sulfate + Magnesium ions + Nitrate ions

SULFIDES - toxic fumes of Hydrogen Sulfide gas (rotten egg gas H_2S) are formed e.g.

Lead Sulfide + Hydrochloric Acid = Hydrogen Sulfide gas + Lead Chloride

CONCLUSIONS:

1. **What was the test and its result for each metal ion (flame test) and for each non-metal group (carbonates etc.)? List or table summary;**
   See above. Students should have these tests recorded in tabular form.

2. **Why was de-ionized water used for making up solutions and dissolving the minerals?**
   To prevent any reaction with mineral ions which may be present in drinking water e.g. Sulfates and Chlorides are common in some water. Teachers may like to test their local water for the class.

3. **What would be the main errors in these tests (be specific for Parts A & B)?**
   Main errors involve contamination and the concentrations of acids and solutions. Students are also capable of mixing up the solutions so bottles should be clearly labelled.

4. **In what situations may a geologist in the field use some of the tests in Part B? Explain.**
   In the field, specimens are often mixed, are in very small sizes and often masked by weathering products e.g. gold and pyrite (fool's gold - iron sulfide) are often confused. A test with acid will show which is the sulfide.

   Similarly, calcite (calcium carbonate) and baryte (barium carbonate) give similar cleavage, colour and rhombic crystals. Whilst baryte has a greater heft than calcite, a flame test (in an ordinary fire) of the powdered minerals will show the green of the barium mineral.

4.4 Answers to Multichoice Questions

Q1.B Q2.B Q3.B Q4.B Q5.C Q6.B Q7.A Q8.D Q9.B Q10.A

4.5 Some Suggestions for the Review Questions

1. **Briefly distinguish between covalent and ionic bonds. Give an example of each in minerals.**
   When some atoms react and combine together, those with only a few electrons in their outer shell or those which only need a few will give up or take in enough electrons to make a stable outer shell and become ions (charged atoms) these then attract together by strong electrostatic forces as ionic bonds. For example, many mineral salts such as halite form crystal lattices this way. Some atoms combine by sharing the electrons in their outer shells which now merge forming covalent bonds e.g. carbon in diamond and graphite.

2. **Why is it difficult to classify minerals by crystal systems?**
   Because it is often difficult to observe a standard crystal shape due to the difficulty in finding larger crystals for all minerals and the multitude of variations in these shapes and intergrowths.

3. **Distinguish between cleavage and fracture.**
   Cleavage occurs when a mineral slits relatively smoothly along natural planes of weakness within their lattice. Fracture is the breaking of a mineral without any relationship to the crystal lattice. When students are looking for cleavage faces by inspection without having to hit the specimen, they may mistake natural growth crystal faces instead e.g. quartz has no cleavage and will fracture in a rough way but it is common to find smooth, flat faces due to the growth of the quartz crystal.

4. **What is chemically important about the element silicon? Why does it form so many compounds?**
   Silicon, like carbon has a valency of 4 i.e. it will form covalent bonds readily with itself and other atoms to give a great variety of possible combinations and so a vast number of Silicate compounds.

5. **Name a mineral which has:**

   a) **hardness 7 and conchoidal fracture**
      Quartz or any of the quartz family.

   b) **hardness 3 and gives carbon dioxide gas when acid is dropped on its surface**
      calcite (dolomite and magnesite both give CO_2 with acid but are slightly harder)

   c) **dark-coloured mineral with hardness of 5 to 6 and two cleavages at about 60^0**
      Hornblende (augite often looks similar but cleaves almost at 90^0 and has short, stubby crystals and is greener in colour)

   d) **heavy heft, cleavage and hardness both 3**
      Baryte superficially resembles calcite but has heavier heft

e) cleavage 2, hardness 6 and is pink in colour
   Orthoclase but it can also be white and confused with plagioclases but close inspection of plagioclase surface will show small lines (or striations) on one cleavage plane.

6. **Give a chemical test for sulfide mineral such as pyrite (fool's gold). How can it be used in a practical way?**
   Addition of an acid will give a faint odour of rotten eggs (hydrogen sulfide). This is useful in the field when both the gold and the pyrite could be in minute specks as dust.

7. **The minerals calcite (calcium carbonate) and barite *(barium sulfate)* can look identical, having the same crystal shape, cleavage, colour and hardness. How could you distinguish between them?**
   As before. Use heft (baryte is heavier) or specific gravity (calcite is 2.7 and baryte is 4.5). Note that baryte is also spelt as barite.

8. **A good specimen of a mineral can be scratched with quartz but it will also scratch apatite. What is the most likely hardness of this new mineral? How could you confirm this conclusion?**
   Quartz has a hardness of 7 so the mineral is < than 7 but apatite has a hardness of 5 so the mineral is harder than this value. 6 would be the best answer but it could be 5.5 to 6.5. One can test this by seeing if the mineral will both be scratched by and also scratch feldspar of hardness 6.

9. **Discuss the importance of ionic substitution in producing a wide range of minerals.**
   Ions of similar size will replace others within the crystal lattice whilst it is being formed (and also later) giving a range of composition. Also, any imbalance in electrostatic charge when this happens will be made up by additional ions.

10. **Research the Internet about the use of minerals in the (a) home, (b) the office and in (c) international trade.** A great variety of uses, this is why the metals industry is so varied and important.

    See:

    https://www.studyread.com/uses-of-metals/

    https://geology.com/metals/

    http://chemistry.tutorvista.com/inorganic-chemistry/metals.html

    http://scitechconnect.elsevier.com/metals-use-medicine/

    http://www.diydoctor.org.uk/projects/metal_used_at_home.htm

4.6 Properties of Common Minerals

Some of the properties of the most common minerals are given as a quick reference on this page and the next:

HARDNESS	COLOUR & STREAK	HABIT	CLEAVAGE or FRACTURE	OTHER PROPERTIES	COMPOSITION	MINERAL NAME
1.0	C = white-green S= white	flakes or aggregates	one perfect	feels greasy, pearly lustre	$Mg_3Si_4O_{10}(OH)_2$	TALC
1.0	C= black-grey S= black	flakes, fibres or aggregates	one perfect	feels greasy, metallic lustre, conducts electricity	C Carbon	GRAPHITE
1.0 to 2.0	C=white S= white	earthy aggregates	Perfect cleavage rarely seen	dull lustre, sticks to tongue	$Al_2Si_2O_5(OH)_4$	KAOLINITE
1.0 to 2.5	C= green S= green	flaky aggregates	one perfect	vitreous lustre	mixture of hydrous aluminium silicates of Fe & Mg	CHLORITE
1.0 to 5.5	C= yellow-brown S= yellow-brown	earthy	no cleavage, conchoidal fracture if hard	dull lustre. opaque	$FeO(OH) \cdot nH_2O$	LIMONITE
2.0	C=white,grey,pink S=white	tabular, fibrous or granular	one perfect, less perfect on other two	silky, pearly of dull lustre, translucent to transparent	$CaSO_4 \cdot 2H_2O$	GYPSUM
2.0 to 2.5	C=colourless, white S=white	Cubic crystals or granular	three perfect at 90°	tastes salty, transparent to translucent	NaCl	HALITE ("Rock Salt")
2.0 to 2.5	C=scarlet red-brown S=scarlet red	Prismatic, rhombohedral or tabular crystals, aggregates or massive	one perfect	heavy heft, dull to adamantine lustre, opaque to translucent	HgS	CINNABAR
2.0 to 3.0	C=silver S=white	Flakes or aggregates	one perfect	Transparent to translucent, vitreous lustre	$KAl_2(AlSi_3O_{10})(FOH)_2$	MUSCOVITE MICA
2.0 to 6.0 (massive) 6.0 to 6.5 (crystals)	C=dark grey-black S=blue-black	Massive but can be fibrous or as aggregates	One perfect with uneven fracture	Metallic to dull lustre, heavy heft	Mainly MnO_2	PYROLUSITE
2.5 to 3.0	C=dark brown-black S=white	Flakes or aggregates	One perfect	Transparent to translusent, vitreous lustre	$K(Mg,Fe)_3 AlSi_3O_{10}(F,OH)_2$	BIOTITE MICA
2.5 to 3.0	C=lead grey S=grey-black	Cubic crystals or granular	three perfect at 90°	Metallic lustre, heavy heft	PbS	GALENA
2.5 to 4.0	C=light to dark green S=green-white	Flakes, fibres or aggregates	No cleavage seen, splintery fracture	Dull to greasy lustre, opaque, feels slippery	Mixture but may be given as $(Mg,Fe,Ni,Al,Zn,Mn)_{2-3}(Si,Al,Fe)_2O_5(OH)_4$.	SERPENTINE
3.0	C=colourless, white or various tints S=white	Rhombohedral crystals	Three perfect but not at right angles	Transparent to translucent, bubbles with acids, double refraction of light if clear	$CaCO_3$	CALCITE
3.0	C=red-grey with irredescent colours S=grey-black	Cubic crystals or aggregates	Indistinct cleavage, conchoidal fracture	Heavy heft, metallic lustre, opaque	Cu_5FeS_4	BORNITE ("Peacock Ore")
3.5 to 4.0	C=white, many tints S=white	Rhombohedral crystals	Three perfect but not at right angles	Transparent to translucent, viteous lustre, bubbles with acid	$CaMg(CO_3)_2$	DOLOMITE
3.5 to 4.0	C=bronze-yellow S=bronze-black	Massive or aggregates	Conchoidal fracture	Metallic lustre, heavy heft	$(Fe,Ni)_9S_8$	PENTLANDITE
3.5 to 4.0	C=brown-black S=brown	aggregates	Perfect cleavage	Opaque to translucent, sub-metallic lustre	ZnS	SPHALERITE

HARDNESS	COLOUR & STREAK	HABIT	CLEAVAGE or FRACTURE	OTHER PROPERTIES	COMPOSITION	MINERAL NAME
3.5 to 4	C=emerald green S=light green	Fibrous aggregates of kidney-shaped lumps	One good in rare crystals, uneven crystals	Vitreous to dull lustre, bubbles with acid	$Cu_2CO_3(OH)_2$	MALACHITE
5.0 to 6.0	C=green-black S=green-grey	Long 6-sided prisms or aggregates	Two good cleavages at 56°	Opaque, vitreous lustre	$(Ca,Na)_{2-3}(Mg,Fe,Al)_5$ $(Al,Si)_8 O_{22}(OH,F)_2$.	HORNBLENDE (an Amphibole)
5.0 to 6.0	C=green-black S=green-grey	Stubby 8-sided prisms or aggregates	Two cleavages at almost 90°	Vitreous to dull lustre, opaque	$(Ca,Na)(Mg,Fe,Al,Ti)$ $(Si,Al)_2O_6$	AUGITE (a Pyroxene)
5.0 to 6.0	C=black, grey-black, green-black S=green or dark brown-grey	Cubic or 8-sided crystals, usually massive or fibrous or kidney shapes	Uneven to conchoidal fracture	Submetallic lustre, heavy heft, radioactive	UO_2	URANINITE
5.0 to 6.5	C=red to grey S=red	Earthy or granules	No cleavage	Metallic to dull lustre	Fe_2O_3	HAEMATITE
5.0 to 7.5	C=blue-white S=white	Long bladed crystals	One perfect and one good	Vitreous lustre	Al_2SiO_5	KYANITE
5.5	C=brown-black S=brown	Massive but rare 8-sided crystals	Uneven fracture	Opaque, metallic lustre	$FeCr_2O_4$	CHROMITE
5.5 to 6.5	C=black S=black	Aggregates and 8-sided crystals	Uneven or conchoidal fracture	Metallic lustre, magnetic	Fe_3O_4	MAGNETITE
6.0	C=white to pink S=white	Tabular or prismatic crystals	One perfect and one good at 90°	Vitreous lustre. Opaque to translusent	$KAlSi_3O_8$	ORTHOCLASE FELDSPAR
6.0 to 6.5	C=white, grey, blue S=white	Tabular or prismatic crystals	Two cleavages at 86°	Vitrous lustre often with striations or blue iridescence	Series grading from $NaAlSi_3O_8$ to $CaAl_2Si_2O_8$)	PLAGIOCLASE FELDSPAR
6.0 to 6.5	C=brass yellow S=green-black	Cubic crystals	No cleavage. Uneven fracture	Metallic lustre, striations on crystal faces	FeS_2 Gives smelly H_2S with acid unlike gold	PYRITE ("Fool's Gold")
6.0 to 7	C=brown-black S=yellow-brown	Stubby 4-faced prisms capped with pyramids	Uneven or conchoidal fracture	Vitreous lustre, heavy heft	SnO_2	CASSITERITE
6.0 to 7.0	C=green S=white-grey	Stubby 6-sided crystals or aggregates	One perfect and another at 115°	Vitreous lustre	$Ca_2Al_2(Fe^{3+};Al)(SiO_4)$ $(Si_2O_7)O(OH)$	EPIDOTE
6.5 to 7.0	C=green S=white	Poor crystals or glassy aggregates	Conchoidal fracture	Transparent to translusent, vitreous lustre	$(Mg, Fe)_2SiO_4$ as a series	OLIVINE
6.5 to 7.5	C=many, mostly red S=white	Crystals with many sides or aggregates	No cleavage, conchoidal fracture	Resinous to vitreous lustre, translucent to opaque	Silicates of Ca, Fe and Al	GARNET
7.0	C=many, often white or clear S=white	Massive or 6-side crystals with pyramids	No cleavage, conchoidal fracture	Translucent to transparent, vitreous lustre	SiO_2	QUARTZ
7.0 to 7.5	C=red-brown S=white-grey	Short prisms often twinned as a cross	Good cleavage in one direction	Opaque, vitreous to dull	$Fe^{2+}_2Al_9O_6(SiO_4)_4(O,OH)_2$	STAUROLITE
9.0	C=many, often black S=white	Tabular or rhombohedral crystals	No cleavage	Vitreous lustre	(Al_2O_3)	CORUNDUM

4.7 Reading List

Anthony, J. W., Bideaux, R. A., Bladh, K. W. & Nichols, Monte C., (Edits). (1995). "Topaz". Handbook of Mineralogy. Chantilly, VA, US: Mineralogical Society of America. ISBN 978-0-9622097-1-0

Barthelmy, D. (2015). Mineralogy Database. (an extremely useful website) http://webmineral.com/

Deer, W.A., Howie, R.A. & Zussman, J. (1992). *An Introduction to the Rock Forming Minerals* (2nd edition ed.). London: Longman ISBN: 0-582-30094-0

Dyar, M.D. & Gunter, M.E. (2008). Mineralogy and Optical Mineralogy. Chantilly, Virginia: Mineralogical Society of America. ISBN: 978-0-939950-81-2.Gem. Institute of America. 2002. *Diamonds and Diamond Grading: The Evolution of Diamond Cutting.* Gemological Institute of America, Carlsbad, California.

Geology.com. (2015). Mineral Uses, Properties & Descriptions. (another great database) http://geology.com/minerals/

Gribble, C.D. & Read, H.H. (1989). *Rutley's Elements of Mineralogy*, 27th Edition. London: Allen & Unwin.560 pages. ISBN: 9780045490110

Haldar, S.K. (2014). *An Introduction to Mineralogy and Petrology*. Atlanta, GA.: Elsevier Science Publishing. ISBN: 9780124081338

Hefferan, K. & O'Brien, J. (2010). *Earth Materials*. NY: Wiley-Blackwell. 608 pages. International Energy Agency (IEA): http://www.iea.org/topics/electricity/

Klein, C. & Philpotts, A. (2012). *Earth Materials – Introduction to Mineralogy and Petrology*. Cambridge University Press. 552 pages. ISBN: 9780521145213

Mayer, W. (1991). *A Field Guide to Australian Rocks, Minerals and Gemstones*. Sydney: Ure Smith.335 pages. ISBN:0 7254 0816 2.

Additional References

http://earthtoleigh.com/documents/worksheets/5.1%20Minerals.pdf
 A simple Multichoice test worksheet on minerals (Open as pdf and Save)

https://www.ebay.com.au/b/Collectable-Mineral-Specimens/3220/bn_2211448 buying minerals

http://www.mineral.org.au/dealers/dealers.html   Australian mineral sites

https://www.mindat.org/ds_8_High-end_Collector_Mineral_Specimens.html International sites for minerals

Chapter 5: Igneous Rocks - The Beginning

5.1 Theme

Rocks are formed from minerals or the components of minerals. Igneous rocks formed from molten magma below the surface of from lava upon the surface and could be considered as the first type of rock and the beginning of the rock cycle

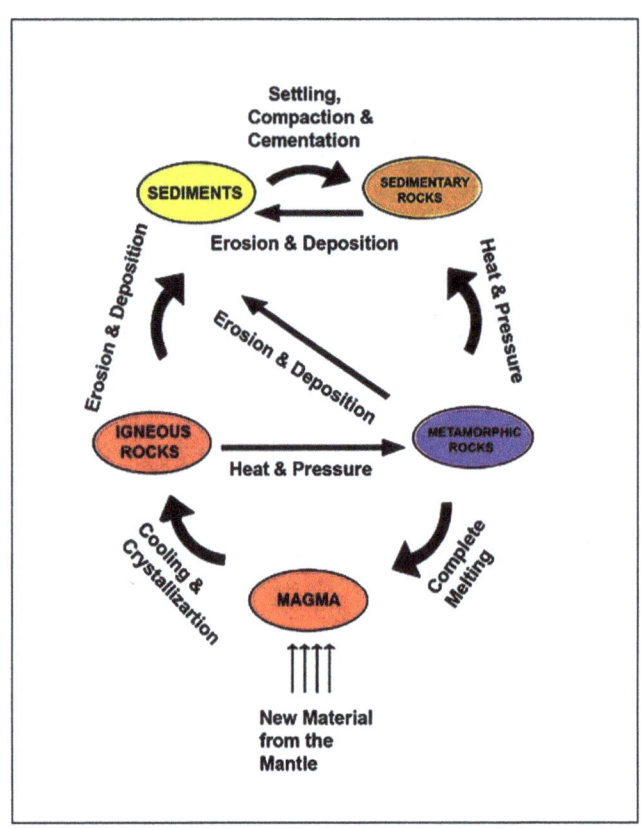

5.2 Rationale

Igneous rocks were the first rocks which formed when the Earth cooled. They are also often crystalline and sometimes show evidence of violent formation. Many of them are very useful as building stones, as bench tops and paving stones. Igneous rocks such as basalt, granite and gabbro are often in great demand by masons and builders. Crystalline rocks such as those which have intruded as batholiths often contain veins of valuable minerals such as gold in quartz and many of the metal sulfides. When these batholiths erode, gold and gemstones may often be found in streams around their margins.

5.3 Notes on the Practical Work (see PRACTICAL MANUAL)

Experiment 5.1: Some Common Igneous Rocks

This experiment reinforces student skills of observation knowledge of minerals and their properties.

A class set of (say) ten of each of the rocks listed (namely granite, gabbro, andesite, trachyte, diorite, rhyolite, basalt, a porphyry, pumice, tuff and a volcanic glass) or similar is recommended. These should be clearly labelled (e.g. R1, R2 or IR01, IR02 etc.) and contained in small boxes (cardboard or plastic) also labelled (Igneous Rock Kit 1 etc.). This list may be reduced or enlarged depending upon local requirements and the institution's collection.

It is hoped that the students will carry over the concepts of mineral identification by careful (hand lens) observation and that smaller crystals relates to faster crystallization (here = cooling). There should also be some classroom connection between these specimen properties and the types of environment (volcanic, shallow intrusive, deep intrusive) in which they are formed and also to some of the most common regions where these rocks are found. If possible, relate these rocks to any found in the local environment. No major detail of volcanoes is required here as they are covered in another chapter.

The appropriate use of media such as photos and video showing some of the rocks in the experiment relating how they are formed would also be appropriate - perhaps after the practical work but before the submission of the practical report (if required).

CONCLUSIONS:

1. In your conclusion, try to name each of each specimen and for each, give a few key words to remember each rock.
Examples may include:

>Pumice could be grey, bubbles
>Granite pink, speckled
>Gabbro dark, crystals
>Trachyte, grey speckled
>Basalt, black uniform

2. Comment on the significance of each rock. i.e. how (and where) it might have been formed. That is, classify each rock as either:

>INTRUSIVE (formed below surface) or EXTRUSIVE (formed on surface) and
>FELSIC light in colour) or MAFIC (dark in colour) e.g. Rock No. # is intrusive, mafic.

These two requirements could be best summarized by:

NAME	KEY WORDS e.g.	CLASSIFICATION
Granite	pink, large crystals	intrusive felsic
Gabbro	dark, large crystals	intrusive mafic
Andesite	green medium crystals	extrusive mafic
Trachyte	grey speckled	extrusive felsic
Diorite	dark medium crystals	intrusive mafic
Rhyolite	light flow lines	extrusive felsic
Basalt	black small crystals	extrusive mafic
Porphyry	big crystals in small crystals	intrusive (mafic or felsic)*
Pumice	grey bubbles floats	extrusive felsic
Tuff	grey ash	extrusive (felsic or mafic)*
Volcanic Glass	smooth shiny	extrusive (felsic or mafic)*

* This depends upon the specimen being used. Many, such as obsidian are of intermediate composition and porphyries can be classified depending upon their main phenocrysts e.g. hornblende porphyry would be mafic but quartz porphyry would be felsic.

There is some difficulty here for some students so perhaps the conclusion of the experiment may be derived collectively after a class discussion. It would be hoped that students would, at least, get some of the answers correct with the less able students being asked the simpler questions (say about the obvious specimens such as minerals in granite etc.).

Experiment 5.2: Igneous Rocks Textures and Composition Exercise.

One of the most important skills used in identification of igneous rocks is the use of thin-section examination through a petrological microscope. A slight rotation of the microscope's stage to get the extinction angle may distinguish between one plagioclase and another to determine whether or not the specimen is a basalt or an andesite and thus two different modes of formation in two different plate tectonic zones. The extinction angle is the angle made by rotation the microscope's stage, using cross-polars, until the plagioclase crystal goes from clear to dark. Textures also tell of the formation of the igneous rock and sometimes the flow direction of any lava.

PART A: Textures
See the following diagram and table:

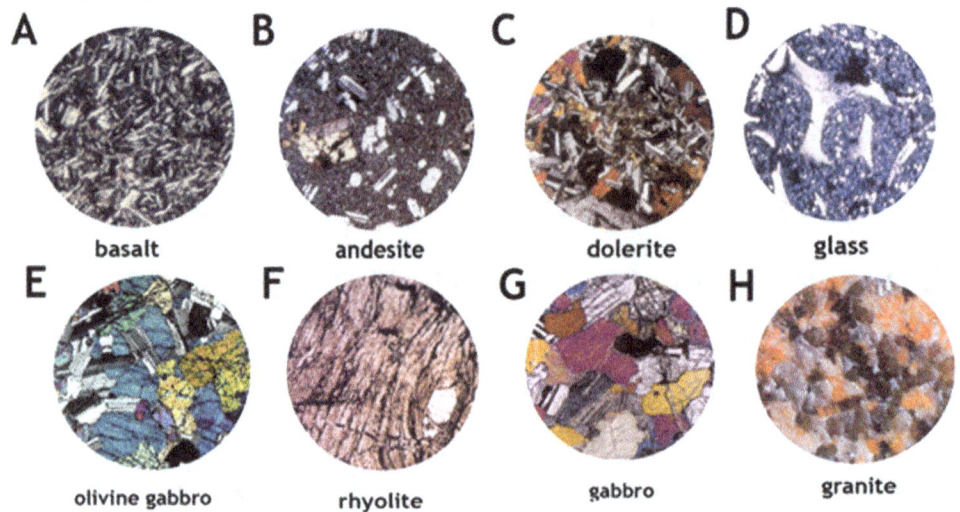

LETTER	NAME	TEXTURE	FORMATION
A	basalt	aphanitic	Lava - hot spots or rifts. Mafic
B	andesite	aphanitic to prophyritic	Lava - subduction zones
C	dolerite	aphanitic to ophitic	Deeper version of basalt - sills, dykes. Mafic
D	glass e.g.obsidian	holohyaline	Lava cooled quickly e.g. in water
E	olivine gabbro	phaneritic to porphyritic	Very deep version of basalt - mid-ocean rifts. Mafic
F	rhyolite	holohyaline with flow lines	Lava and in dykes and sills. Felsic eruptions
G	gabbro	phaneritic	Very deep formation in batholiths etc. Mafic.
H	granite	phaneritic	Very deep formation in batholiths etc. Felsic.

PART B: Point Count:

A good exercise in perseverance and accuracy. The photomicrograph is not a good one but students should be able to distinguish between the large, clear quartz, the grey, clear feldspars, the small, black opaques and the others which would be ferromagnesian minerals. The texture is phaneritic and it is a granodiorite. Some websites showing other photomicrographs in colour are given below. One or more of these could be used in place of the one given but a grid would have to be superimposed e.g. See the USGS site at:

https://pubs.usgs.gov/of/2003/of03-221/htmldocs/thinsect/95mw0107ts.htm

This will give the answer to another version for this experiment:

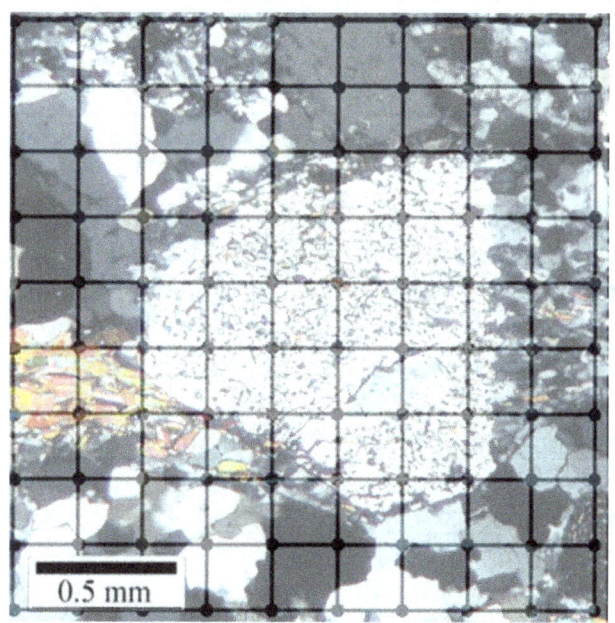

(Photomicrograph of granite courtesy of the USGS)

Alternatively, the following photomicrograph of an olivine gabbro could be used:

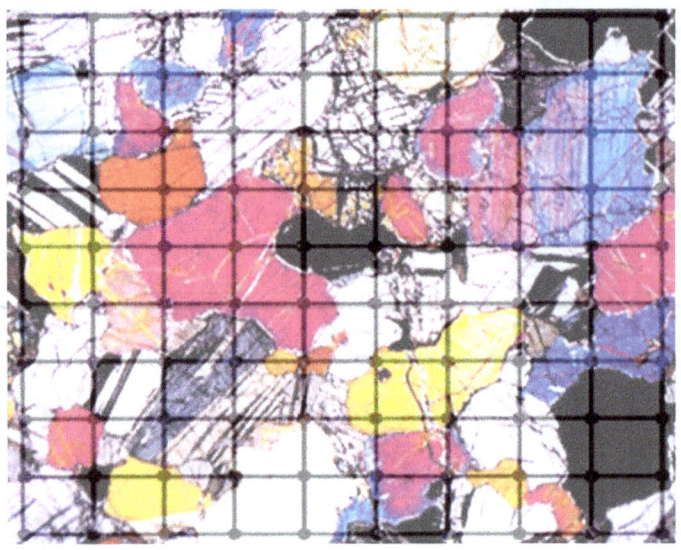

5.4 Other Activities

- **Show some photomicrographs** of some common igneous rocks and relate size and shape to rates of cooling. See:

  https://www.earth.ox.ac.uk/~oesis/micro/

  http://pcwww.liv.ac.uk/geo-oer/wex%20photomicrographs%20-%20igneous.htm

  also show the video from the textbook about how cross-polars work:

   Online Video: A demonstration of cross polars and its effects on thin films. Go to: https://youtu.be/SeQzTVdsvrg

- **Demonstrate the unique property of pumice** by floating a specimen in a beaker of water. Discuss how pumice is formed and why it is often found on beaches.

- **Demonstrate some of the textures** of igneous rocks using fudge (Fun)

  See: https://www.youtube.com/watch?v=mxbmvG5gpAs

  Also: http://www.learnnc.org/lp/pages/3688

- **Make some toffee to simulate volcanic glass (Demonstration)**
  Combine 2 cups caster sugar, ⅔ cup water and 1 tablespoon white vinegar in a metal saucepan (a few drops of red food colouring will add to the effect). Stir over a medium heat until the sugar dissolves completely then immediately stop stirring. Increase the heat to high and bring the toffee syrup to the boil without stirring. Allow the liquid to simmer for 15-20 minutes or until golden. Use a wooden spoon to drag some of the mixture up the sides of the saucepan. This should cool more slowly and give larger crystals. The mix can then be poured into paper cup-cake dishes and allowed to cool to a glassy, transparent sweet. Enjoy!

5.5 Answers to Multichoice Questions

Q1.C   Q2.A   Q3.C   Q4.A   Q5.A   Q6.C   Q7.D   Q8.B   Q9.C   Q10.D

5.6 Some Suggestions for the Review Questions

1. Use Bowen's reaction series to explain the mineral composition of:
    (a) basalt
    (b) granite

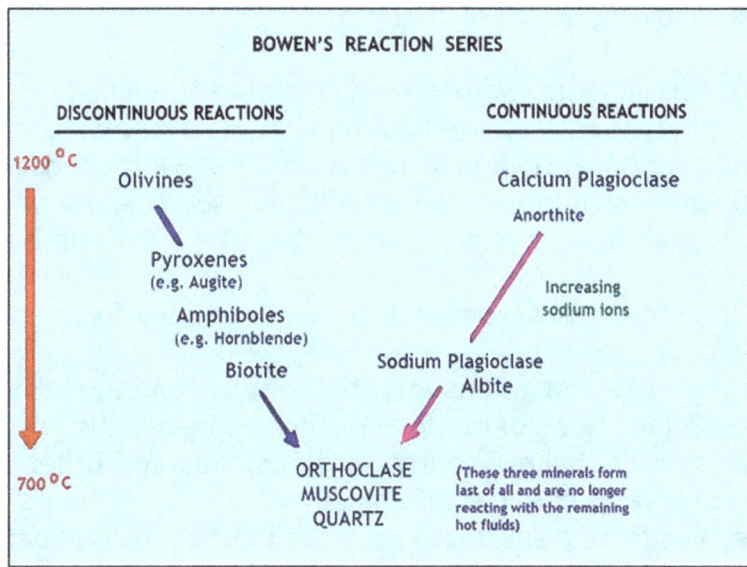

Basalt forms quickly from a high temperature (mafic) lava and so will have those minerals typical of high temperatures such as olivines, pyroxenes and perhaps some amphiboles. Plagioclases will be in the upper range e.g. anorthites.

Granite forms at depth but from a lower temperature magma and will have orthoclase (and perhaps some albite plagioclase) muscovite and quartz forming together and probably a little biotite and amphibole.

2. **Hand specimens of volcanic rocks may be difficult to examine in terms of mineralogy. Explain another way that one can determine the minerals in fine-grained igneous rocks.**
Whole-rock analysis can be done by chemical means using x-ray, gas chromatograph and other techniques to find the elements and their relative compositions within the powdered rock specimen. Knowing the composition and probably the environment of the rock specimen, a mineral composition can be estimated or artificially assigned by comparing the analysis to another known specimen.

   See:
   https://eps.ucsc.edu/research/facilities/geochem-rock.html

   https://brocku.ca/earthsciences/people/gfinn/petrology/chemistry.htm

3. **Explain what is meant by the following terms:**

    (a) **mafic high** concentrations of magnesium and iron minerals
    (b) **discordant structures** intrusions cutting across layers of country rock
    (c) **filter pressing** separation and concentration of denser minerals in a deep magma by the weight of solidifying rock above.
    (d) **accessory minerals** those minerals which only form a small part of the composition because they form last by rapid cooling
    (e) **ferromagnesian minerals** are minerals such as silicates which have high compositions of iron and magnesium (i.e. are mafic).

4. **Define the following terms of texture:**

    (a) **phaneritic** large, well-formed crystals interlocked
    (b) **porphyritic** large, well-formed crystals (phenocrysts) which form first embedded in a finer groundmass (matrix) which suddenly forms last.
    (c) **aphanitic** smaller crystals usually only seen clearly under a microscope
    (d) **holohyaline** no crystals form at all giving a uniform glassy appearance.

5. **What are the modes of occurrence (i.e. where they form and how) of:**

    (a) **granite** deep intrusions (e.g. batholiths) from felsic magma
    (b) **basalt** lava flows of mafic material in shield or fissure volcanoes
    (c) **dolerite** at shallow depths in dykes, sills and other smaller intrusions and often is related to basaltic magma
    (d) **pegmatite** at great depth near the base of granitic batholiths
    (e) **andesite** from the surface eruption of intermediate composition lava.

6. **Which other minerals may be found in an igneous rock in association with:**

    (a) **olivine** any nearby or opposite in Bowen's Series e.g. pyroxenes and anorthite
    (b) **quartz** orthoclase, muscovite mica, albite, biotite
    (c) **anorthite** olivine, bytownite (next plagioclase mineral in the series)
    (d) **biotite** hornblende, albite, quartz
    (e) **augite** labradorite, andesine, hornblende, olivine

7. **Give some uses for igneous rock.**
   Building stone, road metal, paving stones, ornaments, jewellery (obsidian etc.), tools (stone-age axes, arrow heads etc.), source of quartz, feldspar and micas (from pegmatites), abrasives (pumice) etc. see:

   https://www.quora.com/What-are-the-uses-of-igneous-rocks-in-everyday-life

   https://sciencestruck.com/igneous-rock-uses

8. **How could one distinguish between an andesitic and a basaltic lava flow when they look very similar in hand specimen? Explain how the two lava flows could be found together when they come from two different sources.**
Under microscopic examination, thin sections will show that the basalt will contain much calcium plagioclase such as anorthite whereas andesite contains the plagioclase andesine. They can be distinguished by the difference in angle required to make the crystal appear dark under polarised light as the stage is rotated (extinction angle). One such instance occurred in the author's research area where a basalt lava flow had come down a channel eroded in an older andesite layer.

9. **Sometimes the source of extrusive igneous rocks is not obvious due to erosion or distance from the field area. What are some ways by which the source of extrusive rocks can be found?**
Sometimes very difficult and so there is some guesswork involved. One can look further afield using satellite photographs for any igneous structures such as a distant volcanic neck, rounded eroded structure with concentric layers of iron-stained weathering or look for rocks within streams then follow them up or down looking for an increase in igneous rocks within the banks.

10. **Explain how one source of magma in a magma chamber can produce several different types of igneous rocks.**
This occurs because minerals form at different stages of cooling (see Bowen's Reaction Series) and some minerals also sink down through the remaining liquid of the magma chamber or are squeezed out to be concentrated in layers or other deposits.

5.7 Reading List

Geology.com. (2015). *Rocks: Igneous, Metamorphic and Sedimentary*. (Great database) http://geology.com/rocks/

Gill, R. (2007). *Igneous Rocks and Processes – A Practical Guide*. Chichester UK: Wiley-Blackwell. ISBN 9780632063772

Jahns, R.H. (2015). *Igneous rock geology*. Encyclopaedia Britannica. http://www.britannica.com/science/igneous-rock

Jerram, D. (2001). *The Field Description of Igneous Rocks*. Chichester UK: Wiley-Blackwell. ISBN: 9780470022368

Le Maitre, R.W. (Edit.). (2005). *Igneous Rocks: A Classification and Glossary of Terms*. Cambridge University Press. 256 pages. ISBN-10: 0521619483

Marti, J.& Ernst, G. (2005). *Volcanoes and the Environment*. Cambridge University Press. ISBN: 0-521-59254-2.

Michna, P. (2015). *Igneous Rocks*. Earthsci.org.

http://earthsci.org/education/teacher/basicgeol/igneous/igneous.html

Pidwirny M. & Jones, S. (2014). *Characteristics of Igneous Rocks. Fundamentals eBook.* PhysicalGeography.net. Okanaga : University of British Columbia. http://www.physicalgeography.net/fundamentals/10e.html

Scott, P.T. (2017). Rocks - Building the Earth. Brisbane: Felix Publishing. ISBN: 978-0-9946432-2-2

Scrope, G.P. (2009). *Volcanoes: The Character of Their Phenomena, Their Share in the Structure and Composition of the Surface of the Globe, and Their Relation to Its Internal Forces.* University of Michigan Library. 522 pages. ASIN: B002KT3FCY

Additional References

http://www.cas.usf.edu/~jryan/rocks.html

http://www.uta.edu/faculty/mattioli/geol_3313/igneous-textures.pdf

http://www.soest.hawaii.edu/GG/FACULTY/POPP/Sept07_Ch4.pdf

http://www.coconinohighschool-chs.com/lecturenotes/rocks.html

http://www.coconinohighschool-chs.com/lecturenotes/ig%20rock%20table.jpg

http://www.watchknowlearn.org/Category.aspx?CategoryID=2425

http://earthtoleigh.com/documents/worksheets/5.2%20Igneous%20Rocks.pdf
A short Multichoice test on Igneous Rocks (Open as pdf and save)

http://wray.eas.gatech.edu/epmaterials2013/LectureNotes/Lecture17.pdf

Chapter 6: Sedimentary Rocks

6.1 Theme

Sedimentary Rocks are the next type of rock formed from the primal igneous rocks by weathering and erosion. Today, these processes begin as soon as the rock is exposed to the chemicals of the air and water and the effects of gravitation. Sedimentary rocks are also formed as a result of transportation of the weathered and eroded material and then its deposition, cementation and compaction, often in layers.

6.2 Rational

Sedimentary rocks also are valuable as building stones as well acting as reservoirs for oil, natural gas and water. Coal is also a valuable sedimentary rock and many field geologists are employed in searching for this and other sedimentary resources. The nature of the rock and how it was formed often will give indications as to the type of resource which may be found within it.

6.3 Notes on the Practical Work (see PRACTICAL MANUAL 1.)

Experiment 6.1: Some Common Sedimentary Rocks

This experiment shows how sediments will form in layers due to settling of particles (or clasts) and also presents a variety of examples of clastic and non-clastic sedimentary rocks.

A class set of (say) ten of each of the rocks listed; namely conglomerate, sandstone, shale, mudstone, greywacke, coal and limestone or similar is recommended. These should be clearly labelled (e.g. R1, R2 or SR01, SR02 etc.) and contained in small boxes (cardboard or plastic) also labelled (Sedimentary Rock Kit 1 etc.). This list may be reduced or enlarged depending upon local requirements and the institution's collection.

It is hoped that the students will carry over the concepts of mineral identification by careful (hand lens) observation and that larger grains relates to faster settling and that smaller grains relate to slow settling with grain size and composition being good indicators of the sedimentary environment of the rock. Thus, there should also be some classroom connection between these specimen properties and the types of environment. This will be discussed more fully in later chapters but required briefly here to explain the provenance of each rock. These sedimentary environments would include terrestrial, freshwater (lacustrine and fluvial), marine, wind and ice. If possible, relate these rocks to any found in the local environment.

The appropriate use of media such as photos and video showing some of the rocks in the experiment relating how they are formed would also be appropriate – perhaps after the practical work but before the submission of the practical report (if required).

PART A: is entertaining as well as informative. The teacher can prepare bottles of sediment beforehand (and then keep them as a class set for later years) or have the students prepare them at the start of the lesson. If so, each group would need a 2 litre plastic soft-drink bottle, funnel and small samples of pebbly gravel (can be a mix of large and small – between about 1cm to 0.5 cm across), coarse sand, fine sand (preferably of a different colour) and fine silt or dried mud. Only fill the bottle to about one quarter of its height with <u>equal</u> amounts of each sediment. Tightly cap the bottle. Students might like to shake the dry ingredients and see how the sediments settle. Uncap the bottle and fill with water and allow for air to escape (a good observation in itself). It is now ready to be up-ended, shaken vigorously and then quickly stood on the bench for timing the rate of sedimentation. The very fine sediment may take several days and the students usually like to watch what happens to their sediment over time. Even adult students get considerable pleasure out of this simple activity (back to the sandpit, perhaps?)

Students will notice that the gravel will settle instantly and will be mixed with a little sand just as it is found in conglomerates. There may even be a gradual change to the pure sand layer with some conglomeratic sand having mostly sand but a few pebbles. Next there will be a distinct layer of sand or sands if one is lucky enough to obtain two sands which differ in weight (and size). Lastly there will be a layer of mud on top of the sand. The water now will probably be cloudy but in a day or two this will clear and the mud will be covered with a finer layer of silt.

View the online video of this:

Online Video: A short demonstration of the settling of sediment
Go to https://www.youtube.com/watch?v=yaYO4lc_G3M

PART B: Again, with any specimen collection, some students will not achieve perfection and note by observation the exact nature of the grains in the clastic sedimentary rocks but many of the properties will be found and a group discussion could be made after all students have performed their own, individual observations. Estimations of grain size can be done with stereomicroscopes and rulers or just hand lenses for the coarser rocks.

Small dropper bottles of very dilute acid (HCl) can be distributed with trustworthy classes to test for carbonate cement in the clastic rocks (a nil result is also good) and for testing any non-clastic rocks such as the limestone.

CONCLUSIONS:

In your conclusion for PART A, explain why the sediments have settled like this and comment on the time taken for each. Comment on the factors which would control the speed and order of settling.

In general, students should see that the coarser clasts will settle quickly and form a coarse layer at the bottom of the bottle. The sand will settle relatively quickly as a uniform sand layer on top of the gravel and a small layer of larger silt particles may also be seen to settle. The water above will be turgid and by the end of the lesson there should have been some more settling of silt. The bottles may be left over night to see how long the fine sediment will take to settle. This also relates to another experiment layer on turgidity.

For PART B, list the names of the rocks and write two key words to identify the rock and also give a classification for each as clastic (particles can be seen) or non-clastic (no particles e.g. biological or chemical).

Also comment on any internal structures or features seen in each of the rocks and how they and the rock's composition relate to their environment of formation.

Such a conclusion may be something like:

> **conglomerate** – large, rounded pebbles
> **sandstone** – uniform medium-grained
> **shale** – fine, layers
> **mudstone – very fine, uniform texture**
> **greywacke** – medium-grained, grey, semi-rounded clasts
> **coal** – black, dusty, bright bands
> **limestone** – blue-grey, smooth

Shale will be distinguished from mudstone because of its fine layers or lamellae. There may be some fossil leaves in the shale. Coal may be sub-bituminous or bituminous and have some bright bands of vitrain. Limestone may contain fossil corals.

If there is some difficulty here for some students, the conclusion may be derived collectively after a class discussion. It would be hoped that students would, at least, get some of the answers correct with the less able students being asked the simpler questions (say about the obvious specimens such as minerals in sandstone etc.)

Experiment 6.2: Modal Analysis of Sediment

A good experiment to show how some of the classification of rocks can be derived from simple procedures such as sieving. It is a good idea to have an adequate supply of different grades of sediment of each of silt, fine sand, beach sand, coarse sand, fine and coarse gravel. These can be obtained by scouting the local area (and then washed and sieved) or from a local builders' supply yard. The fine gravel used in aquaria is excellent and if possible, obtain some beach sand from between the tidal levels and also high up in the dunes beyond. They can be stored in buckets (e.g. 10 litre) with lids or other suitable containers.

Make sure that the sieves are in correct order and use DRY sediment. Wet sediment sticks to the sieves and also in some sets, the mesh will rust. Encourage accurate measurement of the dry sediment - ask the students why the measuring containers must be tapped on the bench a couple of times when they are being filled (settling of sediment required).

Sieve sets can be purchased online e.g.

http://www.ssapl.com.au/search/results/5a5422b0dd197

https://www.amazon.com/American-Educational-Screen-Sieves-Set/dp/B007F19094

The size of the actual sieve meshes can be marked along the horizontal axis of the column graph. Relate these values with the phi values often seen in the Wentworth Scale, mentioning that values for rock types are obtained this way by dissolving the cement in the rock and lightly hammering it into loose, dry grains.

CONCLUSIONS:

1) **Comment on the sorting for each sand sample;**
   Depends upon the samples of the sediment chosen. Prepared sediments such as those obtained from a builders' supply often are well-sorted. Use a hand lens to check this.

2) **Suggest reasons for any differences or similarities between:**

   1. **Beach and dune sands.**
      Dune sands will be much finer and better sorted as they are blown there from off the beach, often with winds of consistent velocity which will only lift certain sizes.

   2. **River gravel and river sand.**
      An obvious difference in size with the graph showing two main sizes. If poorly sorted, a mixture of the two main show a great variety in sizes from fine to large. Relate this to stream velocity with gravels representing flood/rain sequences.

3. **Beach sand and river sand.**
   This will vary. Probably some river sands may be better sorted than beach sands and some may be finer if the stream is of a consistently moderate flow.

3) **What errors might occur in using volume as a measurement?** Errors due to poor eye level (parallax error) due to the rough surface of dry sediment. Ensure good settling.

4) **Can you suggest a more accurate measurement?** Weight can be used as the main criteria instead of volume.

5) **Comment on the general accuracy and usefulness of this experiment in the study of sedimentary rocks.** This is a good simple experiment to show how <u>clastic</u> sedimentary rocks can be classified using grain size (Wentworth Scale) and how a Modal Analysis can show trends or different levels of sedimentation e.g. a consistent deposition of well-sorted sediments will give a major bar on the graph with few minor fractions.

6.4 Other Activities

- **Make an imitation sedimentary rock** using sand packed into a small cylinder about 2-3 cm diameter. This could be an old cardboard toilet paper roll centre or a piece of cut plastic electrical conduit (about 5-6 cm long). A better idea is a large, plastic syringe with the end completely cut off with a hacksaw (If a class set of syringes cannot be made then one can be used as a demonstration). Small disposable foam cups are also useful.

  1. Place some wet sediment (sand or sand with a few small pebbles for a conglomerate) into the cylinder and then compress it with a plunger (a chemical pestle, appropriately sized piece of dowel stick etc.) or place the sediment into the syringe, place the open end onto a mat and then compress the sediment by pushing down on the plunger.

  2. Push out the plug of compressed sediment onto a mat. Notice that the plug can then easily be broken up into is original sediment i.e. sedimentary rocks need more than just compressed sediment.

  3. Repeat the activity but mix some soluble "wood glue" (any soluble glue is OK provided it is not too thick or dense in colour so that the sediment is hidden by it). Squeeze out the plug as before and leave the plug overnight to dry. If a syringe is used, it must be washed out immediately the plug has been squeezed out.

  4. Discuss the examples of real cements which hold sediments in nature as sedimentary rocks (e.g. silica, calcite, iron oxides).

     Another version of this is found at:

http://www.earthlearningidea.com/PDF/Make_your_own_rock.pdf

- **Cementation of grains** can be seen if one has a projection microscope connected to a video screen. Onto a glass slide, place some very fine sediment so that they are easily seen as separated grains. Now add a drop of concentrated, warm salt water solution. Over a short time, as the water evaporates, the salt crystals will form between the grains. Up end the slide to show the students that they are firmly held together.

- **Pose the question** why salt, a very common mineral in nature, is not a major cement in sedimentary rocks? (too soluble – will easily wash out).

- **Observe that rocks contain gas** (such as air) within their pores and that this gas will flow through the rock by placing various rocks (e.g. granite, quartzite, sandstone) into a large beaker or tank of water. Watch for bubbles. Try submerging pumice (may be much surface air trapped) or vesicular basalt.

6.5 Answers to Multichoice Questions

Q1.D   Q2.B   Q3.C   Q4.C   Q5.A   Q6.C   Q7.A   Q8.D   Q9.A   Q10.D

6.6 Some Suggestions for the Review Questions

1. **Discuss the importance of uplift and mountain building in the rock cycle. What do these processes have to do with sedimentary rocks?**
   Uplift facilitates the processes which can form sedimentary rocks, namely weathering (as the rocks are exposed), erosion and transportation (increased effects of gravity due to height and slope angle), deposition (in newly formed basins) and consequential sedimentation, compaction and cementation.

2. **Define the terms:**
   a. **clastic** rocks which contains visible particles or clasts
   b. **lithification** is the processes which form rocks
   c. **primary structures** are visible shapes formed during sedimentation
   d. **ventifacts** wind-eroded rocks and other shapes

3. **Indicate the past environments which probably produced the following rock types:**
   a. **sandstone** medium-flow river, normal sandy beaches and sand dunes
   b. **shale** slow-flowing rivers, lakes, swamps, deep still oceans
   c. **volcanic breccia** nearby explosive volcanoes
   d. **greywacke** from turbidite currents carrying poorly-sorted sediments down a continental slope into a basin or sea floor.
   e. **fossiliferous limestone** coral reef.

4. **Under a microscope, the grains of quartz in sandstone appear to be formed from smaller grains surrounded with overgrowths of quartz (i.e. it looks like a grain within a grain). What is a probable explanation of this?**
This represents several stages of erosion, sedimentation with cementation. The inner grains represent the first quartz grain which may then be either re-coated with silica or formed into a rock, cemented with silica which coats the grain and then eroded again etc. etc.

5. **An area of rock is found to contain sandstone, sedimentary breccias and evaporites. Indicate the possible environment and give reasons for your answer.**
Several possible e.g. a coastal section near high cliffs and rock platforms. The sandstone comes from the sand which could form as a beach with the inclusion of some breccia material from the collapse of some cliffs above and also some evaporation at low tide of salt etc.

6. **Sandstone can be formed in several environments such as freshwater rivers, on marine beaches and in deserts. Explain how one could tell from which environment a particular sandstone was formed.**
Several ways. Structures such as halite casts and the angle of any ripple patterns as well as marine fossils such as shells will indicate marine conditions. There may also be some conformable marine rocks such as limestones nearby. Freshwater sandstones may also contain coaly fragments and other freshwater fossil remains as well as river structures such as gravel lag deposits. Wind-formed desert sands will have higher angles of repose on any dune or ripple features.

7. **How could a geologist interested in the ancient water or wind flow directions (palaeocurrents) in an area determine the direction of these currents from examining sedimentary rocks?**
He/she would look side on to bedding and note the current bedding formed by water flow and its influence on the original deposition of sediment. Alternatively, the top of bedding planes could be examined for ripple marks. By measuring the directions of the current bedding or the ripples (the current goes in the direction of the longest slope up and the short, steep slope down) for a large number and plotting these as a Rose Diagram graph, one can see the predominant direction of the palaeocurrent.

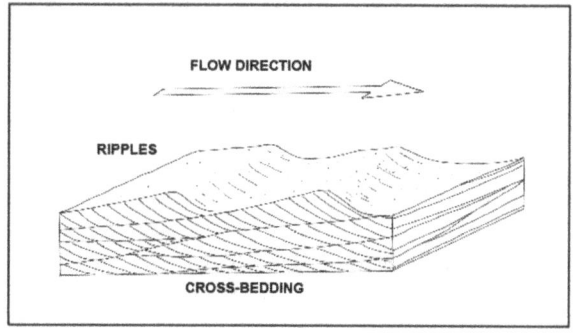

8. **How could this geologist then determine the past velocity of the water or wind flow?**

   This is done by measuring the angle (to the horizontal i.e. the angle of repose) of the cross-section of ripples seen as cross-bedding. A mode or average of such angles taken over a large part of the same bed can be compared to graphs experimentally made with angles and velocities within stream tanks to determine similar rates of motion.

9. **What would be some probable errors in which might occur in attempting to measure the beds within a sedimentary sequence?**

   The main error is in obtaining an exact profile (i.e. side view) of the current bedding. Even finding good, clear current bedding which has not been weathered is an achievement. Ripples with good clarity are even harder to find on exposed, non-eroded bedding planes.

10. **Use the internet to research the arguments for and against the exploitation of coal seam gas in one particular country (USA, Australia, UK or Canada)**

    Do not make generalizations because often the problems with coal seam gas may be different in different countries because the nature of the coal and its sedimentation may be different. Also, there is often a huge polarity in this issue. Many governments and mining companies will state data etc. to show the advantages of CSG whereas environmentalist (and some landowners) will give the opposite. Certainly, there have been some problems in many countries which exploit CSG. The science is often good (in theory) but often this is not properly carried through because of a company's urgency in obtaining and marketing product. Try to obtain a balanced view or several from relatively independent sources (?).

6.7 Reading List

Blatt, H., G. Middleton & R. Murray. (1980). Origin of Sedimentary Rocks. Saddle River, NJ: Prentice-Hall. ISBN: 0-13-642710-3.

Boggs, S., Jr. (2006). Principles of Sedimentology and Stratigraphy (4th ed.). Upper Saddle River, NJ: Pearson Prentice Hall. ISBN: 978-0-13-154728-5.

Collinson, J.D. Thompson, D.B. (1988). Sedimentary Structures (2nd edit.). Boston: Unwin Hyman.

Fichter, L.S. (2000). Sedimentary Rocks. *Department of Geology and Environmental Science James Madison University, Harrisonburg, Virginia.* http://csmres.jmu.edu/geollab/fichter/SedRx/index.html

James, N.P & Jones, B. (2015). Origin of Carbonate Sedimentary Rocks. Chicester, UK: Wiley-Blackwell. 464 pages. ISBN-10: 1118652738

Pettijohn, W. (2005). Sedimentary Rocks. New Delhi: CBS Publications. 220 pages, ISBN-10 812390875X

Prothero, D. R. & Schwab, F. (2004). Sedimentary Geology: An Introduction to Sedimentary Rocks and Stratigraphy (2nd ed.). New York: Freeman. ISBN 978-0-7167-3905-0.

Stow, D. A. V. (2005). Sedimentary Rocks in the Field. Burlington, MA: Academic Press. ISBN 978-1-874545-69-9.

Tucker, M.E. (2001). *Sedimentary Petrology: An Introduction to the Origin of Sedimentary Rocks*. Chichester, UK. Wiley-Blackwemm 272 pages. ISBN-10 1118652738

Additional References

http://earthtoleigh.com/documents/worksheets/5.3%20Sedimentary%20Rocks.pdf
A simple Multichoice test worksheet on sedimentary rocks (answers not included)

http://www.arkansasenergyrocks.com/educators/lessonplans/Roberts-ONGLesson2-Sedimentary-Rocks.pdf (good elementary treatment)

https://www.education.com/science-fair/article/testing-porosity-of-soils/

http://www.geotechdata.info/parameter/soil-porosity.html

http://www.earthscienceeducation.com/taster/Will%20My%20Rock%20Hold%20Water.pdf

https://nptel.ac.in/courses/105103097/27 (permeability and values)

Chapter 7: Metamorphic Rocks

7.1 Theme

Metamorphic rocks are formed from other, pre-existing rocks by the action of heat and/or pressure. A study of these rocks can determine the conditions of these changes and therefore the mechanisms which caused them. Knowing how these changes have taken place, one can also retrace the origin of the source or parent rock.

7.2 Rationale

Metamorphic rocks form from pre-existing rocks by heat and/or pressure and so their nature and mineralogy will indicate what Earth forces have operated in the past. Most are crystalline and are valuable as building material e.g. marble and slate.

7.3 Notes on the Practical Work (see PRACTICAL MANUAL 1.)

Experiment 7.1: Some Common Metamorphic Rocks

This experiment shows some of the features of some common metamorphic rocks and how they can be compared to the original rock formed through the metamorphic process

A class set of (say) ten of each of the rocks listed (e.g. gneiss, hornfels, phyllite, greenstone, slate, marble, schist), named or similarly labelled is recommended. These should be clearly labelled (e.g. R1, R2 or SR01, SR02 etc.) and contained in small boxes (cardboard or plastic) also labelled (Metamorphic Rock Kit 1 etc.). This list may be reduced or enlarged depending upon local requirements and the institution's collection.

It is hoped that the students will carry over the concepts of mineral identification by careful (hand lens) observation and infer that larger grains relate to faster settling and that smaller grains relate to slow settling. Grain size and composition are good indicators of the metamorphic process which formed the rock. Thus, there should also be some classroom connection between these specimen properties and their types of environment. This will be discussed more fully in later chapters but required briefly here to explain the provenance of each rock.

The appropriate use of media such as photos and video showing some of the rocks in the experiment relating how they are formed would also be appropriate – perhaps after the practical work but before the submission of the practical report (if required).

CONCLUSIONS:

This could be similar to:

> **gneiss** – flattened granite.
> **hornfels** – hardened shale, uniform, smooth.
> **phyllite** – smooth, hard, shiny.
> **greenstone** – green, smooth, slippery.
> **slate** – flat, hard, brittle layers.
> **marble** – white, granular or smooth, hard.
> **schist** – shiny, flakes.

7.4 Other Activities

- **Effect of heat.** Mix a little flour (coarse if available) with a little water and make a soft dough. Place it into a small aluminium dish (a small pie dish etc.) and heat it gently over a Bunsen burner (or use an oven if convenient), noting the changes. Eventually it will become hard, shiny and water proof.

7.5 Answers to Multichoice Questions

Q1.B   Q2.A   Q3.D   Q4.A   Q5.C   Q6.B   Q7.B   Q8.D   Q9.D   Q10.B

7.6 Some Suggestions for the Review Questions

1. **Define the term texture. List the major textures found in metamorphic rocks.**
   Texture refers to the way that different minerals are formed and orientated within the rock. Main textures are: hornfelsic, granoblastic, slaty, phyllitic, schistose and gneissic.

2. **Define each of the following:**

   (a) **porphyroblastic** - a texture in which large crystals (porphyroblasts or phenocrysts) are embedded in a finer crystalline groundmass
   (b) **augen** – from the German for eye, these are oval shaped concentrations of a mineral within a gneiss.
   (c) **diagenesis** – refers to the changes which occur as rocks form from sediment or some previously existing rock.
   (d) **metamorphic aureole** – is the area around a hot body (e.g. batholith or dyke/sill) in which metamorphism of the country rock has occurred.
   (e) **facies** – is the character of a rock expressed by its formation, composition, and fossil content.
   (f) **protolith** – from the Greek meaning first rock, it is the original, unmetamorphosed rock from which a given metamorphic rock is formed

3. **What is a polymorph? Give an example and explain its significance in the identification of metamorphic rocks.**

   This is the ability of a solid material to exist in more than one form or crystal structure e.g. aluminum silicate (Al_2SiO_5) can exist as the minerals kyanite, sillimanite and andalusite depending upon the temperature and pressure.

4. **What is meant by the term retrograde metamorphism? How could you distinguish a retrograde metamorphic rock from one which had been formed normally (progressively) to that same grade?**

   Retrograde metamorphism is the mineralogical adjustment of relatively high-grade metamorphic rocks to temperatures lower than those of their initial metamorphism and is uncommon. The term *diaphthoresis* (from the Greek to degrade) has also been applied to this change. Distinguishing between the two may be difficult (some Internet research here), but the presence of garnets may suggest a high temperature to low temperature regression.

5. **Give the possible parent rocks for the following metamorphic rocks:**

   a) **quartzite** - sandstones, arkoses and similar.
   b) **marble** - limestones, dolomites.
   c) **calc-silicate hornfels** - from impure limestones and dolomites and other calcium rich sediments which also contain clays and sand.
   d) **hornfels with the minerals hypersthene and hornblende** – rocks of basaltic composition at higher temperature.
   e) **amphibolite** – from mafic rocks such as basalts
   f) **gneiss** - from high pressure on many rocks e.g. granite

6. **Account for the pairing of regional and contact metamorphic belts. Use a diagram to explain your answer.**

   Usually the folding and dynamics which causes regional metamorphism will also cause igneous intrusions which will give contact metamorphism.

7. **Outline the metamorphic rocks which would be formed from a granite intrusion into dolomite. How could a petrologist distinguish between these rocks and those rocks formed by a similar intrusion into limestone?**

   This intrusion could produce calc-silicate hornfels as well as marble whereas limestone will only give marble (with the mineral calcite as its main constituent)

8. **In the field, what would be some indicators that the local area is of general metamorphic origin?**

   There may be physical indicators such as the hardness of the rock e.g. marbles and quartzites but more often considerable foliation with regional metamorphic rocks especially the presence of micas, often seen in waterways as shiny flakes. Other metamorphic minerals such as garnets may also be found there.

9. **Gneiss and granite have the same basic minerals - quartz, orthoclase, plagioclase, biotite and muscovite. Given each of these rock specimens in the laboratory, how could you distinguish between them?**

   These minerals in gneiss are very typically flattened and have the same orientation, whereas these minerals in granite have random orientation.

10. **Use the Internet to research about the:**

    (a) metamorphic rocks in your immediate location
    (b) economic use of metamorphic rocks or materials in them

    Slates, marbles and many others are well known as building stones with marble also being used for ornamentation and statues. Some dolomitic marbles are used as sources of magnesium and some regional rocks are sources of gems such as garnets. See:

https://www.reference.com/science/metamorphic-rocks-used-d2513eb0a144d55a

https://prezi.com/a58v4ladkh86/metamorphic-rocks/

https://socratic.org/questions/what-are-some-examples-of-uses-for-metamorphic-rocks

7.7 Reading List

Blatt, Harvey and Robert J. Tracy, *Petrology*, W.H. Freeman, 2nd ed., (1996). p.355 ISBN 0-7167-2438-3

Frost, B.R. & Frost, C.D. (2013). *Essentials of Igneous and Metamorphic Petrology*. Ambridge University Press. ISBN-10 1107696291

Gillen, Cornelius. (1982). *Metamorphic Geology: an Introduction to Tectonic and Metamorphic Processes*, London; Boston: G. Allen & Unwin ISBN 978-0045510580.

Marshak, Stephen. (2009). *Essentials of Geology*, W. W. Norton & Company, 3rd ed. ISBN 978-0393196566

USGS, (2014). *US Geology in the Parks: Metamorphic Rocks*.
http://geomaps.wr.usgs.gov/parks/rxmin/rock3.html

Vernon, Ronald Holden. (2008). *Principles of Metamorphic Petrology*, Cambridge University Press ISBN 978-0521871785

Wicander R. & Munroe J. (2005). Essentials of Geology. Cengage Learning. pp. 174-177. ISBN 9780495013655.

Winter, J.D. (2009). *Principles of Igneous and Metamorphic Petrology*. New Jersey: Prentice Hall. 720 pages. ISBN-10 0321592573

Additional References

https://www.msnucleus.org/membership/html/jh/earth/metamorphic/jhmetamorphic.pdf

https://www.nps.gov/para/learn/education/upload/para%20lesson4.pdf

http://newyorkscienceteacher.com/sci/files/topic-media.php?media=Worksheet&subject=earth+science&subtopic=rocks+and+minerals

http://www.d.umn.edu/~mille066/Teaching/2312/2016%20Labs/Lab9-Meta%20hand%20samples.pdf

https://en.wikibooks.org/wiki/High_School_Earth_Science/Metamorphic_Rocks

Chapter 8: Weathering and Erosion

8.1 Theme

As soon as the minerals within rocks are exposed to the gases, solutions and temperature changes within and above the Earth's surface they begin to change physically and chemically. Physical weathering is usually a simple change into smaller particles but chemical weathering often involves more complex changes as the minerals turn into new substances which then can be further worn down or carried away.

8.2 Rational

The surface of the Earth and its materials are constantly being worn away by chemical and physical interaction often resulting in the large-scale removal of material down slope. Engineers, construction workers and those who use the land should be mindful as to the instability of the subsurface rock and of sloping surface when constructing roads, dams, bridges and large buildings. Farmers and graziers should have a good understanding of the potential for land and soil erosion so that their properties and crops can be maintained.

8.3 Notes on the Practical Work (see PRACTICAL MANUAL 1.)

Experiment 8.1: Weathering

This experiment relies on previous knowledge of basic minerals and their identifiable features. Students should be aware of the basic minerals within granite such as quartz, orthoclase and plagioclase feldspars, muscovite and biotite micas and hornblende. They should also have studied the iron oxides such as haematite and limonite and clay (kaolin) as these are the main constituents seen in granite and weathered granite. Students should also know that limestone and marble have calcite as their main constituent.

The identification of the crystals seen in the fresh granite specimen is good revision and may be done as a class revision with less able classes but otherwise should be able to identify these minerals themselves. Some students may have difficulty as the minerals may be masked by the iron oxides.

Care must be taken with the heating of small, fresh granite specimens as rocks often explode when strongly heated. Dilute acid will also hard the eyes. Eye protection and aprons are necessary and if possible strong gloves to hold the tongs. If the class is limited in their ability, these experiments may be done as a demonstration.

CONCLUSIONS:

These experiments should show how rocks can be broken down by physical and chemical processes.

1. **What was learnt in each of the three sections about weathering:**

   PART A: Weathering of granite
   This shows how the quartz grains remain relatively unchanged, perhaps slightly broken, but the other minerals have altered. The feldspars and muscovite have become clays and probably mostly leached out. The mafic minerals such as the biotite, augite and hornblende have also become clays but also there has been a considerable change to red iron oxides.

   PART B: Physical weathering of granite.
   Simply shows that sudden changes in heat will break the rock up into smaller pieces.

   PART C: Chemical weathering of limestone or marble
   This shows how such carbonate rocks dissolve in acid solutions such as water charged with carbon dioxide (i.e. carbonic acid). The rock will completely dissolve and carbon dioxide gas is liberated. This can be related to structures such as caverns in Karst erosion.

2. **Give the word equations for any chemical changes:**

   - Feldspars ⟶ Clay Minerals ⟶ Bauxite + Solutions
     (aluminium silicates of potassium, sodium and calcium) (hydrated aluminium silicates) (complex aluminium oxide-hydroxide) (metal salts etc.)

   - Micas ⟶ Clay Minerals + Iron Oxides + Solutions
     (silicates possibly with possibly aluminium, potassium, iron) (hydrated aluminium silicates) (if iron was in the mica)

   - Ferromagnesian Minerals ⟶ Clay Minerals + Iron Oxides + Solutions
     (iron & magnesium silicates) e.g. hornblende augite, biotite (hydrated aluminium silicates) (e.g. red haematite, yellow limonite)

   - Quartz ⟶ No change but will physically breakdown into sand

   - Limestone (or marble) ⟶ calcium hydrogen carbonate solution + carbon dioxide gas

3. **Relate the artificial, laboratory actions to the real source of change for the actions observed in the experiment.**
   The laboratory experiment has provided an extremely faster rate of physical and chemical change compared to the slow, but methodical changes which go on in nature. These are the changes in heat of day and night and the action of very weak acidic and still natural waters over a very long time.

4. **Discuss any errors which may have prevented the appropriate observations from being made. Suggest any improvements.**
   Usually this experiment works well but the sudden breakup of granite when heated may distract from the experiment. If the acid is too weak, the reaction by be too slow to observe the dissolving of the solid piece of rock.

Experiment 8.2: Stream Table Experiments

There are many possibilities here. Apart from the system shown in the experiment, one can use an old fish tank or make long, thin sedimentation tanks out of thick, transparent acrylic (Perspex) to show how sediments settle from a side-view perspective. This is also useful to show water ripples (use sand on the bottom, shallow water and make waves on top by pushing with a piece of rectangular wood) and sand dune formation (use fine sand and blow gently with an old hair dryer).

Coarse sand is useful in this experiment although a mixture of silt, gravel and sand is better to show differences in erosion and settling. As with all wet experiments, some care is needed in setting the apparatus up and also at the finishing stages of cleaning. It is recommended that one has a laboratory set of uniform-sized (use sieves for this) washed gravel (say 0.25 to 0.5 cm diameter), coarse sand, medium sand and washed silt kept in 10 Litre buckets with lids. At the end of the activity, students would be expected to empty their trays full of water and sediment into waste buckets for the appropriate sediment or mixed sediment. All can be dried later on the trays used in the experiment. The mixed sediments would then have to be reclaimed using a sieve set.

Students must ensure that no sediment goes down the sink into the plumbing system. Also ensure that excessive water is not used so that the receiving trays overflow onto the desk. This experiment may be modified with trays having an overflow tube near their top so that pure water can drain into the sink.

CONCLUSIONS:

1. What are the main features of rivers seen in this simulation?
Students should be able to observe the formation of meanders, possibly some point and longitudinal bars and some braided stream formation at the end of the main channel.

2. What happens to the width of the stream's course as it moves downslope?
The stream channel should get wider and perhaps degenerate into an alluvial fan with braided streams.

3. How is the sediment in the stream carried?
By the action of the flowing water due to the downhill pull of gravity. It will also take the pathway of least resistance.

4. Describe the motion of the sediment at the (a) start and (b) end of the river
At the start the water flow is faster, more direct and less sinuous but eventually widens, slows down and is more meandering.

5. Discuss the use of such models (as in this experiment) in the study of real river environments.
These models are used in real scientific studies often on a much larger scale. They are most useful because they allow for a variety of controls and parameters such as change in slope, soil type and surface features. More on this later.

6. Why was the sand moistened before the experiment?
This stops the absorption of water into the dry soil – infiltration. It also prevents any severe initial erosion of the dry, loose sand at the very start of the activity. Also, with steeper slopes the dry sand will move down the tray.

7. What errors could limit the use of such a model? Give some examples from this experiment.
If the sand is not wet enough it will slide rather than stay in place and will also be eroded at too high a rate. If the sand is too wet it will also slip down the tray and not show any erosion too clearly. If the water flow is too fast, erosion will occur too quickly and not show the effects of downstream patterns.

8. How such an experiment could be improved.
Some practice is required to get the sand in the right wetness consistency and the water flow sufficient to show the appropriate erosion features of fast and slow streams. Also, the angle needs to be just right for this water flow. It could be improved by using a larger model but this may be impractical.

Experiment 8.3: Soil Testing Experiments

There is an overlap here with any rural or agriculture classes. This experiment is a good introduction to agronomy and other rural sciences.

A laboratory set of soils (local and imported) is recommended. Artificial loam can be made from equal amounts of powdered clay, fine silt and sand. Use a mortar and pestle to grind up dried clay. Many of these materials can be purchased at hardware stores or builders/gardening supply yards.

PART A: Measuring the Acidity (pH) of the soil.

The concept of pH was first introduced by the Danish chemist Søren Peder Lauritz Sørensen at the Carlsberg Laboratory in 1909 and revised to the modern pH in 1924 to accommodate definitions and measurements in terms of electrochemical cells. The notation pH generally stands for the power of the hydrogen ion (although often using other languages but still referring to power) or in scientific terms minus the logarithm of the hydrogen ion concentration. As these values are very small (with negative indices values), the minus sign allows an easy whole number reference where acid is from 0 to 6 (part values and values less than zero occur), neutral is 7 and alkaline is from 8 to 14.

Universal Indicator is typically composed of water, propan-1-ol, phenolphthalein sodium salt, sodium hydroxide, methyl red, bromothymol blue monosodium salt, and thymol blue monosodium salt. The colours that indicate the pH of a solution, with Universal Indicator are:

Colour	pH range	Description
RED	<3.0	strong acid
YELLOW	3.0 to 6.0	weak acid
GREEN	7.0	neutral
BLUE	8.0 - 11.0	weak alkali
PURPLE	>11.0	strong alkali

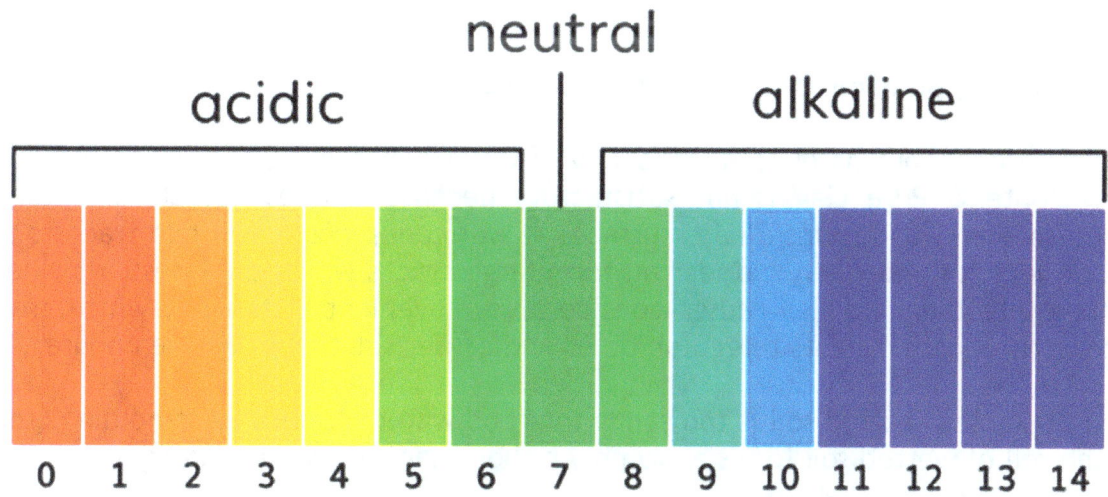

No colour? See:

https://www.shutterstock.com/search/%22ph+chart%22

Note that Universal Indicator is a mixture of dyes and will stain the skin and other materials. Skin will give an acidic reaction!

Litmus paper can also be used (red=acid, blue=alkaline) but this is only approximate and not recommended. Litmus is a mixture of lichen species, particularly *Roccella tinctoria* and the word *litmus* is literally from coloured moss in Old Norse. Red cabbage leaves and geranium flowers can also be used to indicate acidity.

Soil pH kits are available from some garden shops and plant nurseries, but these kits are only accurate to half a pH unit. They measure pH in water, whereas laboratories measure pH in calcium chloride ($CaCl_2$). The pH in calcium chloride gives values about 0.5 to 0.8 lower than pH in water. More expensive kits, costing about $500, measure pH in calcium chloride to an accuracy of about 0.2 pH units. Measuring pH using simple classroom indicators (i.e. in water) would be adequate to show the basic principles of soil acidity. In practice, if:

- pH is above 6: soil acidity is not a problem yet.
- pH is below 6: there is a risk of soil acidity problems, so methods that will reduce the rate of acidification such as adding lime should be applied.

Acidity is a potential problem on any farm, and its prevention should be part of your soil management plan. Strongly acid soils (pH less than 5.5) require special care. Adding large amounts of lime can result in potassium and magnesium deficiencies and in sandy soils, over-liming can cause deficiencies in trace elements such as zinc, manganese and iron. Plants affected by acidity also become more prone to disease and plant growth is slow and lateral root growth is stunted. Acid soil also limits nodulation and nitrogen fixation by legumes; however, potatoes, sweet potatoes and watermelons tolerate some acidity.

PART B: Clay and Soil -the Farmers' Technique

This is a quick but hardly scientific way of determining if a soil has a satisfactory clay content. Pure clay is not satisfactory because it does not allow water to penetrate and also forms a hard surface. In wet conditions it will bog tractors and other vehicles and when it dries will form large cracks which then destroys wheels falling into them. It also causes considerable movement in the soil which cracks open brick houses, slants telegraph poles and causes slanting in wooden houses.

Too little clay and the soil is too sandy for good water retention and holding crops. If the roll of wetted soil forms a crumbly ribbon then this is satisfactory:

If the ribbon measures less than 2.5 cm long before breaking, it is **loam** or **silt**.

If the ribbon measures 2.5 to 5.0 cm long before breaking, it is **clay loam**.

PART C: The Effect of Lime on clay soils

Clay soil is made up of millions of tiny particles, giving it a very fine texture. This has advantages and disadvantages. On the plus side, clay soils are rich in nutrients (the particles provide multiple surfaces where nutrients can stick) and they hold water well. However, they are also prone to compaction, waterlogging, and can be sticky when wet and tough when dry. These problems are made worse if a clay soil is cultivated when wet.

Soils with excessive clay may be improved by:

- adding organic matter such as well-rotted manures, mulch and compost. This binds the particles together, improves the soil structure and allows water and oxygen to move through the soil more easily.

- adding sand to clay soils may help improve drainage and aeration, but one would need to dig in huge quantities to have any real effect.

- addition of gypsum (often sold as clay breaker) can also be beneficial and again this can be dug through the soil. It is important only to cultivate fine-textured soils when they are 'damp-dry', not when they are wet.

- adding garden lime to the soil will increase the soil's pH bringing it back into equilibrium. At this point, all soil elements including phosphorus, potassium and calcium should become freely available to plant's roots. This is the ideal combination for beneficial bacteria and fungi.

The addition of lime in the form of calcium carbonate acts to consume hydrogen ions contributing to acidity as well as removing thick cation (+ ions) layers around clay particles ensuring structural stability of the soil. The hydrogen ions in acid soils have a disastrous influence on plant growth this can be reduced when the carbonate portion of lime reacts with the soil. Oxygen associated with carbonate binds to the hydrogen ions located in the soil to form water. It does so by temporarily forming hydrogen carbonate which then transforms into water and carbon dioxide. Therefore, the majority of hydrogen ions are converted into water and no longer pose a threat to plant health.

An additional role of lime is the improvement in soil structure and aggregate stability. The calcium portion of lime displaces substances such as hydrogen and aluminium located around the clay particles. Calcium has a smaller molecular weight in comparison to these substances and the area between the clay particles is therefore reduced. The advantage of this occurring is that a reduction in space results in a smaller distance between clay particles and this creates weak bonds of attraction called dipole forces. This binds clay together and although the dipole forces may be very weak, there are quite a number of these attractions overall contributing to make strong cohesive bonds. These strong cohesive bonds form aggregates and ensure that the soil does not deteriorate into its individual sand, silt and clay components. This is the case with dispersion, which is not favourable for plant growth.

PART D: Soil Components and Texture

The textural designation of a soil is determined by its relative portions of sand, silt, and clay particles, and indicates which of the three most influence the soil's properties. Sand, silt, and clay soil properties are obviously dominated by those respective fractions. For example, clay soils (generally more than 40 percent clay) are often poorly drained. On the other hand, well-drained loam soils are mixtures of sand, silt, and clay in roughly equal proportions, and are well drained. A sandy loam, however, has much more sand and much less clay than does a clay loam.

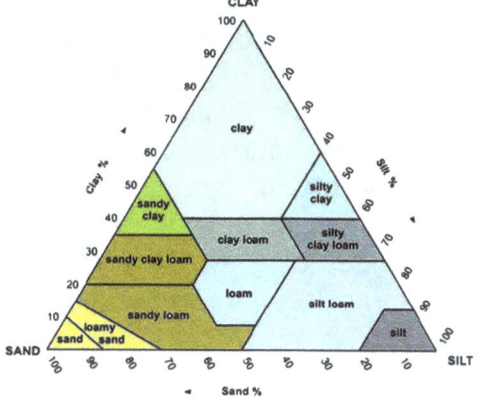

CONCLUSIONS:

1. Write a full conclusion with discussion about the observations made during this experiment.
As appropriate to the soils being used.

2. What are the main factors (a) shown in this experiment and (b) from additional research which are important in understanding about the usefulness of soils?
Soils should have a good texture with balanced clay, sand and silt (as well as mineral and organic nutrients).

3. What is pH?
This refers to the acidity of the soil. See:

http://soilquality.org.au/factsheets/soil-acidity

https://www.qld.gov.au/environment/land/management/soil/soil-properties/ph-levels

https://www.dpi.nsw.gov.au/__data/assets/pdf_file/0003/167187/soil-ph.pdf

https://www.agric.wa.gov.au/soil-ph-and-plant-health

4. What is the original source of (a) clay and (b) the organic matter found in some soils?
Clay comes from the breakdown of felsic minerals in rocks such as feldspars and micas. The organic matter comes from the decomposition of vegetable matter which has grown in the soil or transported there as well as animal manure. This is often re-worked by worms and other organisms.

5. Are there any improvements or additions which should be made in this experiment?
A wider variety of different soils could be used and there could be a microscopic examination of each soil using a binocular microscope. In addition, the moisture content could be tested using cobalt chloride paper.

8.4 Other Activities

- If convenient, **organise an excursion to an old local cemetery** and measure the depth of indentation of the dates on older gravestones. Graph the depth against the year on the gravestone. Older stones should have more weathering and so more shallow indentations. Choose the same rock of each gravestone (e.g. marble or granite). There should be an inverse relationship between depth of indentation and year. One could repeat this activity using a different set of gravestone rock to compare the different rates of weathering between (say) granite and marble.

See:
https://www.ucl.ac.uk/earth-sciences/impact/geology/london/citycemetery/weathering

https://www.wired.com/2016/10/want-tombstone-last-forever-make-quartzite/

https://www.livescience.com/18681-cemetery-gravestones-weathering-acid-rain.html

- **Students could do a photographic survey of the weathering of statues** and building stones in their area. For this and the previous activity, students could think of the local climatic conditions which may increase weathering. Are there other examples of weathering here (e.g. by tree roots etc.)? See:

  https://sciencing.com/weathering-affect-monuments-4324.html

  https://link.springer.com/article/10.1007/s12665-010-0826-6

- **Freezing water and expansion.** Place a known volume of water into a disposable cup, marking the water level with a pen. Place into a freezer overnight and note the increase in volume. Have the students explain why this is so and give some common examples (e.g. water pipes bursting when frozen, car radiators breaking due to ice expansion) or repeat this with small measuring cylinders accurately noting the increase in volume. Calculate this increase as a percentage by volume (should be around 9% due to the formation of the crystal).

- **Make ice blocks** in the usual way using a large compartmented tray or use many small plastic containers but add some coarse gravel (say with pebbles about 0.5 to 1.0 cm in diameter). When hardened, have the students grind some sandstone or other friable rock with the ice to simulate how the till within glaciers grinds country rock.

- **Demonstrate a turbidity** current by pouring a very liquid slurry of mud down a ramp within a glass tank (e.g. an old fish tank). See:

  https://www.youtube.com/watch?v=8gYJJjxY8g0

  https://www.youtube.com/watch?v=ZhDQnnONWl4

  https://www.geolsoc.org.uk/ks3/gsl/education/resources/rockcycle/page3660.html

8.5 Answers to Multichoice Questions

Q1.B Q2.A Q3.B Q4.D Q5.B Q6.C Q7.C Q8.A Q9.D Q10.D

8.6 Some Suggestions for the Review Questions

1. Discuss the major similarities and differences between:

   a. **weathering and erosion**
   Weathering is the chemical and/or physical change to minerals in rocks and usually occurs on site.

   b. **chemical weathering and physical weathering**
   Chemical weathering involves some irreversible change in which new substances (compounds) are produced whereas physical erosion does not; only breakdown to smaller sizes and shapes occur.

   c. **hydration and hydrolysis**
   Hydration is the simple addition of water to a compound whereas hydrolysis involves the breaking of chemical bonds within a compound.

   d. **off-loading and exfoliation**
   Off-loading is the cracking of crystalline rock into parallel sheets due to internal stresses as weight is removed from above. Exfoliation is a more general term (some also include off-loading here) as a crystalline rock peels off in sheets from its outer surface. This may be due to expansion and contraction caused by heat changes.

2. **What are some methods used to prevent soil erosion in mountainous areas?**
Soil erosion can be prevented by good agricultural methods such as contour ploughing (ploughing around hillsides, not down them), the use of terraces and stone walls.

3. With the aid of a diagram, explain the nature and formation of a:

   a. **scree slope** -steep pile of loose rock fallen down from above
   b. **turbidite** - sediment or rock formed from a sudden underwater flow of debris off the continental shelf
   c. **tor** - rounded boulder of crystalline rock formed by exfoliation.
   d. **soil profile** - side view of a sample a soil on site showing its layers. Refer to appropriate photos or diagrams in the textbook.

4. List some of the weathered products, such as minerals, solutions etc. of each of the following rocks:

   (a) **granite** - quartz, clay minerals, iron oxides, resistant mica and solutions of cations.
   (b) **basalt** - mostly clays and iron oxides with solutions
   (c) **limestone** - solutions of calcium hydrogen carbonate and some magnesium hydrogen carbonate

(d) quartz sandstone - only smaller pieces of quartz

5. **What is meant by thixotropic clay? What is its significance in engineering geology?**

   Some thixotropic clays and ash deposits may be considered as extremely slow-flowing liquids and appear to be solid but if shaken or vibrated (by volcanic activity or man-made explosions) they suddenly become liquid and will easily flow causing landslides and some volcanic mudflows (some lahars). Construction of bridges or dams on such clays can be a problem if the clay is vibrated.

6. **Building a chalet or home in high mountainous areas poses some additional problems to constructing similar buildings on the lowlands. Discuss.**

   Mountain constructions are subject to landslides due to the instability of the ground, through erosion by sudden surface flows of water and by avalanches of snow. There are additional problems involving the access and use of materials in high, steep regions.

7. **What is the significance of the development of cyanobacteria 2100 million years ago?**

   As blue-green alga, these organisms absorbed carbon dioxide from the atmosphere (the major gas at this time) and converted it to oxygen, thus beginning the development of the Earth's oxygenated atmosphere.

8. **In granitic mountains, streams sometimes contain small specks which reflect light like gold. On closer inspection, these are not metallic pieces of gold but thin, transparent flakes. Suggest what they might be and how they got there.**

   Often a problem for new prospectors! These are more resistant flakes of muscovite mica produced as granite chemically weathers. In sunlight, their single cleavage plane reflects the light and looks like gold.

9. **Research the soil types in your area. What are they used for mainly? What are some of the problems associated with these soils?**

   This answer requires individual research findings. Try local councils, Department of Agriculture offices and conservation groups.

10. **Use the Internet to research hydroponics. What is it? Why is it used?**

    Hydroponics is the science of growing plants without soil under controlled conditions. Because the appropriate nutrients, temperature and light are constantly available, plants thus grown will develop better and faster. Moreover, plants can be grown indoors or in glasshouses (even in a schoolroom window) in regions with harsh climates.

8.7 Reading List

Allaby, M. (2000). *Basics of Environmental Science*. London: Routledge. 340 pp. ISBN-10: 041521176X

Chesworth, Ward, (edit). (2008). Encyclopaedia of Soil Science. Dordrecht, Netherlands: Springer. ISBN: 1-4020-3994-8

Conway, A., R. Reynolds, R., Dise, N., Dubbin, B. & M. Gagan, M. (2006). *Air and Earth*. Milton Keynes, UK: Open University. 326 pp. ISBN-10 074926988X

Drever J.I. (2005). Surface and Ground Water, Weathering, and Soils: Treatise on Geochemistry, Second Edition, Volume 5. Elsevier, Atlanta, GA 30303, USA. 644 pp. ISBN 0080547591, 9780080547596

Goudie, A.S., Viles H. (2008). Weathering Processes and Forms. In Burt T.P., Chorley R.J., Brunsden D., Cox N.J. & Goudie A.S. Quaternary and Recent Processes and Forms. Landforms or the Development of Geomorphology. 4. Geological Society. pp. 129-164. ISBN 1-86239-249-8.

Skiba, S. (2014). *Physical (Mechanical) Weathering of Soil Parent Material*. In Gliński,J., Horabik, J. & Lipiec, J. (Edits.). *Encyclopedia of Agrophysics*. Springer International, Dordrecht, The Netherlands. pp 595-596. ISBN: 978-90-481-3584-4

Yerima, B.P. K. & Van Ranst, E. (2005). *Introduction to Soil Science: Soils of the Tropics*. 2005. Bloomington, Indiana. 440 pp. ISBN-10: 1412058538

Additional References

http://www.wyorksgeologytrust.org/misc/Teaching%20weathering%20in%20West%20Yorkshire.pdf

https://www.pbslearningmedia.org/resource/nat08.earth.geol.eros.lpbreakit/breaking-it-down-weathering-and-erosion/#.WneayXxLfX4   (EXCELLENT!)

https://undsci.berkeley.edu/lessons/pdfs/teacher_station_list.pdf

http://science-class.net/archive/science-class/Geology/weathering_erosion.htm

http://www.soils4teachers.org/lessons-and-activities#General9 (EXCELLENT!)

https://www.nrcs.usda.gov/wps/portal/nrcs/detail/soils/edu/?cid=nrcs142p2_054308

https://jameskennedymonash.wordpress.com/2014/09/15/colourful-chemistry-chemistry-of-universal-indicator/

http://cwfs.org.au/2017/01/09/chemical-structural-benefits-lime-soil/

https://www.rhs.org.uk/advice/profile?PID=144

https://www.dpi.nsw.gov.au/__data/assets/pdf_file/0006/127257/Does-my-soil-need-lime.pdf

http://www.biology-resources.com/documents/Experiments/12%20Soil/Soil%20experiments.pdf

Chapter 9: Environments of Weathering and Erosion

9.1 Theme

Landforms are the result of the combination of surface building processes such as uplift and igneous activity as well as destructive processes such as weathering and erosion. The nature of these processes can often be deduced by observing the landforms seen in an area. Such processes continue to operate today.

9.2 Rationale

The appearance of the land and many landscapes which are considered beautiful are a result of the geological events of the past. The processes which formed them can still be observed today, this is the Principle of Uniformitarianism. A good understanding of these processes will enhance the students' appreciation of the beauty of such landscapes as well as assisting in working out the geological history of a region. This chapter is also useful to geographers who are concerned with the surface features and their uses.

9.3 Notes on the Practical Work (see PRACTICAL MANUAL 1.)

Experiment 9.1: Karst Simulation

This experiment is a simple model to show how limestone and similar carbonate rocks are eroded underground to form caves. Whilst it is suitable for younger students, this model works well to show the basic principles.

Students should pour in sufficient water so that the level of water in the cup comes up to about halfway into the first sugar cube layer. This simulates the water table in nature, below which there is considerable reaction and formation of bigger spaces.

This experiment can be deleted and a more dramatic demonstration can be performed using a bigger container with flat sides (say, a small fish tank) and a lot more sugar.

CONCLUSIONS:

1) **Where does most of the reaction take place during the initial stages?**
   Down the vertical edges of the cubes, then along the horizontal joints.

2) **Is this consistent with what occurs in nature? Why?**
   This is consistent with nature as the limestone usually has many joints formed by horizontal bedding planes and also by vertical and oblique joints formed when the rock is compacted.

3) **What happens at the base of the cubes where water has collected (i.e. below the water table of this model)?**
   More of the sugar dissolves due to the increased surface exposure below the water. If sufficient of the sugar dissolves, there also may be subsidence, simulating what actually occurs. If the surface is made from a different type of porous soil, then sinkholes on the surface may form with this collapse.

4) **What errors could limit the use of such a model?**
   Sugar may dissolve too quickly and not allow suitable time for observation. If so use very cold water.

5) **How such an experiment could be improved.**
   Try it on a larger scale with more sugar cubes pushed more closely together as bigger cubes.

 Online Video: Venture with the author underground into a limestone cave. Go to https://youtu.be/OCbccRGRh84

Other Activities for this experiment:

- **Assign different temperatures** of water to class groups to see if temperature of the water affects the dissolving of their limestone (sugar).

- **Use calcium carbonate** (as in the limestone) by making a thick paste in water then pasting it onto a large (say 20 cm x 20 cm) glass or rigid plastic sheet. When almost dry, scrape a brickwork pattern across the slab then place another, identical glass sheet over the top making a sandwich of calcium carbonate brickwork between transparent sheets. Stand the sandwich vertically (and clamp it) in a suitable tray and then slowly and carefully add some dilute acid (vinegar or dil. HCl) to the top.

- **Make stalactites** by making a saturated (i.e. no more solid will dissolve) solution of any of sodium bicarbonate, salt or Epsom salts. Fill two identical jars or cups (say 250 ml) and place them both into a tray about 15-20 cm apart. Obtain a length of absorbent string or twine and fold it in half and then in half again to make an M shape. Hang both ends into each of the cups so that the centre of the twine hangs down between the cups well below half way (but with sufficient distance from the tray). Allow to stand for a few days.

  One might have different groups placing their experiment in different locations of the classroom where there are differences in air currents or temperature. This activity could also be done at home and the length of the stalactite measured at the same time each day to graph the rate of growth.

- **Invite a member of a local Caving (Spelaeology/Spelunking) club** to give a talk about their activities (most states have an association of clubs. Refer to the Internet)

- **If convenient, run an excursion** to a local developed cave region (guided tour)
- **Use the Internet to show videos about limestone caves.** Some resources would be:

https://www.youtube.com/watch?v=PpbxFpAZmSQ (animation showing development)

https://www.youtube.com/watch?v=dB-4qX9zFVM (introduction to caving)

https://www.youtube.com/watch?v=9MoWGyrzBUw (caving and creatures)

https://www.youtube.com/watch?v=IKZ9EHewjZg (New Mexico - BBC video)

https://en.wikipedia.org/wiki/Caving_organizations (world caving associations)

http://www.caves.org.au/ (Australian Spelaeological Association)

Experiment: 9.2: The Shape of Rivers

This experiment will require some teacher pre-experimentation to ensure that the bed is initially at the right angle to get appropriate meanders. 2.5 degrees worked for the author but the type of sediment used (medium-coarse sand) will also play apart as will the application of the water at the headwater end.

CONCLUSIONS:

1. **Is there any relationship between the slope of the river bed and its sinuosity? Explain and give reasons.**
   With increased slope, there should be a <u>decrease</u> in Sinuosity Index as the speed of the water will cut a more direct channel preventing the development of meanders.

2. **Discuss the use of such models (as in this experiment) in the study of real river environments. Are there other factors which would determine the sinuosity of a river?**
   Such models are often used on a large scale by researchers who often go to considerable trouble to make the model realistic. The stream table may be many tens of metres long. Other factors which would determine the sinuosity of a river will include the type of rock or sediment of the valley, vegetation, speed of the river due to its source, the confining nature of the river valley and any tributaries which may be entering.

3. **What errors could limit the use of such a model?**
   Any of the factors mentioned above if different from the model, especially the flow rate and the local geology.

4. **How such an experiment could be improved?**
   Attempt to make the model as close to reality as possible. In this case. The addition of water should be gradual (e.g. use a small watering can) and have a wide tray.

Online Video: Take a dangerous trek down the 4000 metre deep Colca Canyon, western Peru.
Go to https://youtu.be/PwRYBlTmWZw

Experiment 9.3: Stereopairs and Landforms

This is an old-fashioned technique going back to the early days of aerial reconnaissance but it is still useful for the students to see relief in three dimensions and to understand the need for such capability. Today, 3D modelling is very advanced and surface and sub-surface features can be represented in detail, rotated and analysed. Moreover, 3D satellite imagery can make use of different wavelengths to obtain data from different aspects of the terrain or ocean.

If possible, it would be better if the students had a set of aerial or satellite stereo pairs which would be of good clarity that they can use with their stereo viewers to overlap the photos and move their viewers around. A few good sets of prominent landforms such as volcanoes, faults, tilted beds etc. would be useful. These are often purchased from local State land offices. Stereo pairs photos and viewers can also be purchased from:

http://www.ascscientific.com/stereos.html (books etc. at a good price)

http://www.aerialsearch.net/contact.html

https://www.amazon.com/American-Educational-Aerial-Photographs-Individual/dp/B005QDR1MG   (Stereo book)

https://www.amazon.com/s/ref=nb_sb_noss_1?url=search-alias%3Dindustrial&field-keywords=stereoscopes   (Stereoscopes)

The photographs in the Practical Manual (reproduced from those at NASA/JPL and USGS who provide photos free of copyright) are low resolution and come from the following sources respectively:

PHOTO 1:
This shows part of the San Andreas Fault in California taken from the Space Shuttle northwest of Los Angles in 2000. The details are at

https://photojournal.jpl.nasa.gov/catalog/PIA02776

Students should be able to notice the linear feature (a fault line) running diagonally up the photo separating the areas of mountainous relief from the plains. There also is another linear feature running across the photo near its top, cutting the main fault at about 60 degrees.

Students and staff with computers can access the original stereo pair at:
https://photojournal.jpl.nasa.gov/jpegMod/PIA02776_modest.jpg

and attempt to see the 3D image without using a viewer, but many find this difficult.

PHOTO 2:
This shows the centre of Mt. St. Helens sometime after the 1980 eruption, it is an aerial view taken obliquely by the US Geological Survey. It shows the development of a new central lava dome.

The original stereo pair comes from:
http://extra.listverse.com/amazon/stereograms/mountsainthelens.jpg

CONCLUSIONS:

1) **Comment of the use of satellite and lower altitude aerial photos to locate and describe landforms.**

   These provide detail that normal surface mapping along set traverses cannot show. Using more professional stereoscopes, data such as heights and lengths can be determined. As many of these photos have a vertical exaggeration, even small-scale surface features stand out. Moreover, photos can be taken using different wavelengths so that specific features can be seen.

2) **What errors would limit the use of stereo pairs?**

   Whilst good detail is provided, the causes of such features and detailed lithology is often not possible. A detailed ground survey is needed.

3) **How such a technique be improved?**

   Use of specific frequencies and better high resolution cameras can be used. Satellite photos can be supported by low level stereo photography (as in Photo 2).

Some resources for the use of **stereo pairs** include:
http://nzphoto.tripod.com/sterea/stereoview.htm

http://www.seos-project.eu/modules/3d-models/3d-models-c02-p04-s01.html

http://www.stereoscopy.com/faq/index.html (good for a wide range of DIY techniques and contacts)

http://www.aerialsearch.net/StereoViewer.html

https://www.e-education.psu.edu/natureofgeoinfo/book/export/html/1817 (more complex techniques to identify and scale features)

Other satellite stereo pairs which can be viewed on computers can be found by searching images under stereo pairs and include:

- Honolulu (NASA/JPL)
  https://photojournal.jpl.nasa.gov/jpegMod/PIA02738_modest.jpg

- Pasadena (NASA/JPL/NIMA) – a low altitude oblique angle view
  https://ia800203.us.archive.org/9/items/VE-IMG-347/PIA02737.jpg

- Fiji (NASA/JPL)
  https://photojournal.jpl.nasa.gov/jpegMod/PIA02785_modest.jpg

- African Volcano (NASA/JPL/NIMA)
  https://photojournal.jpl.nasa.gov/jpegMod/PIA03357_modest.jpg

- Wellington (NASA/JPL/NIMA)
  https://photojournal.jpl.nasa.gov/jpegMod/PIA02749_modest.jpg

All of these provide coloured stereo pairs which do not require stereoscopes. They can be downloaded and then made into separate Left/Right images and overlapped (electronically or printed) and then viewed with a stereoscope, but the SIZE of the presented image is critical. Some experimentation is needed.

Anaglyphs are another 19th century technology which has been greatly improved. These are images with overlapping red/cyan (or green) outlines. When viewed using glasses with red (left) and green (right) lenses, the image stands out in three dimensions. These have not been included in the printed Practical Book for costs purposes and are best viewed on computer screens (enlarged is great!).

A class set of anaglyph glasses can be easily made from cardboard and red and green cellophane. A site giving a template for the glasses is given below. This may be printed onto thick paper or cardboard and then the lenses of cellophane glued onto the appropriate sides so that the RED is LEFT and the GREEN (cyan if you can get it) is on the RIGHT:

https://stereo.gsfc.nasa.gov/classroom/glasses.shtml

Alternatively, professionally-made glasses can be cheaply purchased at

https://www.ebay.com.au/i/222885418168?chn=ps&displtem=1

Some excellent examples of landform anaglyphs can be found at the USGS site on US National Parks and then clicking on the THUMBNAIL button for each park:

https://3dparks.wr.usgs.gov/#

Also, with time and care one can make one's own anaglyphs using programs like Photoshop of places visited by the school and other interesting photos of landforms or specimens for viewing at home or for sending 3D photos to remote or absent students. This is done by making two copies of the photo as different layers, and then changing the Blending Option to give each a red or green aspect. See:

http://www.vicgi.com/anaglyph-images-photoshop.html

http://scecinfo.usc.edu/geowall/makeanaglyph.html

https://www.diyphotography.net/create-3d-anaglyph-images/

One can also download 3D anaglyph movies, there are many from NASA but more on that in later chapters:

https://www.youtube.com/watch?v=GTNCULSFi-4
(Iceland.glaciers and rivers)

Computer generated 3D with rotation is used often with surface and subsurface features but this requires the appropriate software and good computer memory.

https://upload.wikimedia.org/wikipedia/commons/1/1d/Sierra_Blanca_Peak_%28New_Mexico%29_3D_version_1.gif   Sierra Blanca Peak (New Mexico)

9.4 Other Activities

- **Do the VIRTUAL RIVERS activity** using computers in class or at home (as a project/exercise)

    http://www.sciencecourseware.com/VirtualRiver/

- **Take a mini-excursion** to a local river (or major drain). Sample and test the water and measure stream flow by measuring the time taken for a float to travel a measured distance.

- **Make models of landforms** using topographical maps and tracing major contours onto corrugated cardboard or sheets of foam and then gluing the layers together. These can then be painted.

    http://www.3dgeography.co.uk/making-3d-maps

- **Conduct a quiz** (or even a marked test) by projecting images of specific landforms associated with erosional/depositional environments. These can be downloaded from the Internet. e.g.

    KARST - stalagmites, stalactites, columns, Rillenkarren, helectites, shawls.
    RIVERS - meanders, point bars, longitudinal bars, oxbow lakes (billabongs), terraces.
    COASTLINES - imbricated beach, sea stack, rock platform, tessellated pavement.
    ALPINE - crevasse, horn, arête, seracs, cirque lake, esker, drumlin, erratic.
    DESERT - pedestal rock, yardang, dreikanters, barchan, mesa, inselberg

- **Revise River environments** by playing the video of Iguassu Falls from the textbook:

Online Video: Visit the Iguassu Falls (Iguazú Falls) on the Brazil-Argentine border. Travel by high speed boat into the base of the falls and visit the Devil's Throat.
Go to https://www.youtube.com/watch?v=qycwdJz9BJc

Questions (for the students):

Where are these falls located? – on the Iguazu River on the border of the Argentine province of Misiones and the Brazilian state of Paraná. A little upstream is Paraguay.

What is significant about these falls? – one of the widest falls in the world.

How long is the edge of the falls? - 2.7-kilometres.

What are their greatest height? – several cataracts in steps varying in height from 60 and 82 m.

What is the rock type of the plateau over which the water falls? Basalt

- **Revise Glacier environments** by playing the video from the textbook:

Online Video: Trek across glaciers in Switzerland and New Zealand at https://youtu.be/zo1bQLLArIM

Questions (for the students):

How do glaciers form? – from the compaction of snow in high snowfields then the slow movement of this ice downhill.

What are the hazards associated with exploring glaciers? - crevasses, moulins (holes with water rushing down them), slippery slopes, rockfalls from surrounding hills, collapses sending water downstream.

What importance do glaciers have? – provide fresh drinking water to many communities adjacent to mountains. Recreational climbing. Albedo effect.

What evidence is there for the passage of glaciers in countries where they no longer exist? - many glacial U-shaped valleys, striations on rocks, erratic boulders mid-valley.

- **Use the Internet to find out where the major deserts** of the world are located. How did they form? What is desertification? Are their places today which may become deserts in time?

  See: http://www.antarcticglaciers.org/glaciers-and-climate/glacier-recession/mapping-worlds-glaciers/

https://nsidc.org/cryosphere/glaciers/questions/located.html

https://www.gtav.asn.au/documents/item/615

http://www.grid.unep.ch/glaciers/pdfs/3.pdf

9.5 Answers to Multichoice Questions

Q1.C Q2.C Q3.A Q4.C Q5.B Q6.A Q7.D Q8.A Q9.C Q10.A

9.6 Some Suggestions for the Review Questions

1. What might the typical landscape structures to be found in:

    a. **limestone country** - limestone pinnacles, gorges, sinkholes, caves.
    b. **glaciated plains** - erratics, drumlines, eskers, glacial lakes.
    c. **coastal regions** - beaches, sea-cliffs, sea-stacks, tombolos.
    d. **high alpine regions** - peaks, glaciers, cols, U-shaped valleys.

2. Discuss the importance of water in the processes of forming landscape. Water is soft, so how can it shape rock?
   Water is a major agent in forming river and coastal regions as well as other landscapes mostly because of the tumbling and abrasive action of loose material carried in the water. Hydraulic force of water can also widen some weakened cracks in rock and remove loose regolith.

3. Discuss the major similarities and differences in regard to processes, landforms and sediments of:

    a. **weathering and erosion**
       Weathering is the breakdown of rock on site due to changes in heat and crystal form, whereas erosion is the total removal of soil and rock.

    b. **river erosion and coastal erosion**
       River erosion and coastal erosion both remove rock by the action of particles in moving water with river erosion caused by a continual flow downhill. Coastal erosion is caused by regular motion of incoming waves.

    c. **water erosion and ice erosion**
       Both remove rock by the grinding effect of particles within the medium and whereas water erosion involves a rolling action of these particles within water, glacial erosion involves the grinding by rock embedded within the ice.

d. water erosion and wind erosion
   Both are caused by the effects of particles grinding rock with wind erosion having smaller, sharper particles which grind at the base of most rock.

4. **Discuss the importance of climate on weathering and erosion.**
   Temperature and humidity are the main factors here. A hot, wet climate will cause more chemical weathering and erosion by moving water whereas a cold, dry climate will tend towards physical weathering and erosion other means such as ice and/or wind. Hot, dry climates also favour physical weathering and erosion by wind and water when it does rain.

5. **With the aid of a diagram, explain the nature and formation of:**

   a. scree slope
   b. levees
   c. medial moraine
   d. tombolo
   e. zeta-curved beach
   f. yardang

   Please refer to the appropriate diagrams in the textbook

6. **List some of the environmental structures likely to be formed from the following rock types:**

   a. **Granite** -rounded boulders, tall, steep-sided hills
   b. **Basalt** extensive flat areas. Especially on heights
   c. **Limestone** – tall, sharp pinnacles, sinkholes.
   d. **quartz sandstone** – dissected plateaux, layered cliffs

7. **Discuss the preparation and equipment needed to do research into the flow of water deep inside a cave within a karst area several hundred kilometres from home.**
   This would require an expedition of several days so as well as the usual caving equipment (helmets, boots, overalls, lamps, ropes, flexible ladders), one would need a tough vehicle, camping equipment and food and water for the duration. Specialized flow-rate instruments may be used although a simple float, stopwatch and measuring tape would also suffice.

8. **What is a spelaeothem? Explain the factors which may cause the different types of these structure.**
   Spelaeothems are the deposited calcite structures found within limestone caves due to the dripping or flow of calcium hydrogen carbonate solution which then gives calcite crystal as the water evaporates. The type of spelaeothem depends upon the nature of the water flow (and therefore the rainfall above), the surface over which it may flow and the evaporation rate within the cave due to natural air currents. Stalactites

form as water drips slowly from the ceiling whereas stalagmites and rimstone pools form as the water accumulates on the cave floor. Columns form when stalactites and stalagmites join together and shawls are formed as water trickles down the walls. Other dripstone formations depend upon the slope and nature of the surface of the sides or floor of the cave.

9. **What are the major hazards associated with exploring a glacier?**
Glaciers are dangerous places. The ice can be slippery if it has little covering of snow or moraine and negotiating seracs, crevasses and moulins is extremely difficult, requiring pitons, ropes, rigid ladders. There is also the chance of exposure due to the cold, damp conditions as well as falling ice and boulders from surrounding cliffs.

10. **Playa lakes are most commonly found in desert areas and are important sources of minerals.**

    (a) **How are these minerals formed?**
    They are evaporites formed when minerals are leached from local mountains by rare rain storms. In some places, such as in the Andes, this leaching is also supplemented by the upwelling of mineral-rich volcanic water.

    (b) **Give some of the uses of some of these minerals.**
    These dried lakes are rich sources of nitrates for fertilizers and explosives, gypsum for building, salt and lithium minerals for batteries.

See: https://pubs.usgs.gov/gip/deserts/minerals/

9.7 Reading List

Allaby, M. (2000). *Basics of Environmental Science*. London: Routledge. 340 pp. ISBN-10: 041521176X.

Anderson, R.S. & Anderson, S.P. (2010). *Geomorphology: The Mechanics and Chemistry of Landscapes*. Cambridge University Press. 651 pp. ISBN-10: 0521519780

Charlton, R. (2013). *Fundamentals of Fluvial Geomorphology*. London: Routledge. 320 pp. ISBN-10: 0415505518.

Conway, A., R. Reynolds, R., Dise, N., Dubbin, B. & M. Gagan, M. (2006). *Air and Earth*. Milton Keynes, UK: Open University. 326 pp. ISBN-10 074926988X.

Guinness, P. & Nagle, G. (2009). *IGCSE Geography*. London: Hodder Education. 216 pp. ISBN-10: 0340975016.

Holden, J. (2011). *Physical geography: The Basics*. London: Routledge. 176 pp. ISBN-10: 0415559308.

Huggett, J. (2007). *Fundamentals of Geomorphology*. London: Routledge. 480 pp.
ISBN-10: 0415390842.

Migon, P (Edit.). (2010). *Geomorphological Landscapes of the World*. New York: Springer. 375 pp. ISBN-10: 9048130549.

Walthan, T. (1974). *Caves*. London: MacMillan. 238 pp. SBN 333 17414 3

Willett, S. (Edit.). (2006). *Tectonic, Climate, And Landscape Evolution (Geological Society of America Special Papers)*. Geological Society of America. 434 pp. ISBN-10: 0813723981.

Additional References

https://www.nps.gov/ozar/learn/nature/cave.htm

https://www.waitomocaves.com/education-centre/education-news/

https://www.nationalgeographic.com/news-features/son-doong-cave/2/#s=pano1072

https://www.pbslearningmedia.org/resource/ess05.sci.ess.earthsys.virtmap/virtual-cave/#.WrSGv39LfX4

https://www.handsontheland.org/teachers/data/caves-and-karst-nps-2017.pdf

http://www.earthlearningidea.com/PDF/185_Sink_hole.pdf

https://www.youtube.com/watch?v=PpbxFpAZmSQ (Good Cave animation)

https://pmm.nasa.gov/education/sites/default/files/lesson_plan_files/Models%20of%20Land%20and%20Water%20TG.pdf

http://www.ga.gov.au/education/classroom-resources

https://en.wikibooks.org/wiki/High_School_Earth_Science/Introduction_to_Earth%27s_Surface

https://en.wikibooks.org/w/index.php?search=landforms&title=Special:Search&go=Go&searchToken=nn4vvygx28noepsmsmbvab1x

https://en.wikibooks.org/wiki/High_School_Earth_Science/Glacial_Erosion_and_Deposition

https://en.wikibooks.org/wiki/Historical_Geology/Deltas

Chapter 10: The Hydrosphere - Waters of the Earth

10.1 Theme

The unique properties of water and the ideal position of the Earth from the Sun and of the planet's size enables it to have a hydrosphere which covers over 70% of its surface. Oceanographic studies of the sea floor have led to the discovery that it was produced by spreading from oceanic rifting centres and destroyed at subduction zones.

10.2 Rationale

The hydrosphere is the main feature of Earth - the blue planet. A knowledge of the properties of water and how its several forms exist on the planet enables a better understanding of the Earth's other spheres and how life depends upon water.

10.3 Notes on the Practical Work (see PRACTICAL MANUAL 1.)

Experiment 10.1: Properties of Water

Water is a very interesting substance. Its polarity causes some interesting effects and this experiment should give the students an entertaining insight into the properties of water which is a major part of our lives.

PART A: Polarity of Water

The plastic rod becomes positively charged by friction when rubbed and when placed near a stream of water will attract the negative ends of the water molecules. The water stream should be set at the slowest rate which will give a fine, regular stream. Water turbulence will distract from the effect.

The experiment will not work very well on a humid day because the water in the air will discharge the rod. If this is the case, demonstrate the effect by charging the rod in front of a heat lamp to remove air moisture from the area around the rod and quickly place it near the water stream.

PART B: Surface Tension

Because of the polar bonding of water within a drop or below a water surface, the water molecules are pulling inwards due to their attraction. However, at a surface, this pull only comes from below and so gives a surface tension which stretches the surface with enough force to float light-weight flat objects and allows various insects to walk on water. Adding some detergent causes a disruption to these polar bonds and the surface tension breaks down and the paperclip will sink.

PART C: Cohesion and Adhesion

Water molecules will stick together, cohesion, and to other polar or neutral surfaces, adhesion, because of the nature of the water molecule's polarity. Some surfaces will allow water to adhere, hydrophilic substances, and other substances will repel water, hydrophobic substances. A glass surface, thoroughly cleaned and dried, will allow water to adhere to it and so the water stream poured onto the glass very slowly will break up and spread across the glass. The wax on waxed paper is hydrophobic and so there is not attraction to the water molecules. Here the internal forces of cohesion are greater than that of adhesion to the wax paper and so the water stays in a ball or flattened ball and quickly rolls down the paper. This can be repeated using ordinary paper and the water will adhere and be absorbed by the paper.

When a thin capillary tube is placed in water, the forces of adhesion to the glass are stronger than the cohesive forces so the water moves up the tube. A little food dye or potassium permanganate to colour the water gives an improved effect.

PART D: Capillary Action

When the forces of water adhesion are greater than the forces of cohesion, water will move up a substance against gravity as in Part C. This is capillary action and it is a very useful scientific tool. If the marking pen is a mixture of dyes, these will be separated out just like the scientific tool of Paper Chromatography. A little experimentation will give marking pens which have ink made from several dyes which have different molecular weights. This determines how they will rise up the paper. Lighter-weight dyes will climb higher. The ink must also be soluble in water so that the rising water column will carry the ink. Students should experiment by making dots near the base of the paper using different pens on the same spot.

PART E: Heat of Vaporization

Heat is needed to break apart the polar bonds of the water molecules to separate them when liquid water turns to a gas. A draft of air will help to physically remove some of the surface molecules and thus speed up the evaporation process and the removal of heat of vaporization. Alcohol (e.g. ethanol C_2H_5OH which is less polar), because of its molecular nature evaporates faster and in doing so absorbs more heat which comes from the finger. Ether works better but is more toxic.

PART F: Solubility and miscibility

Solvents tend to dissolve solutes which are similar e.g. polar solvents such as water will dissolve substances which are polar or have charged ions e.g. salt crystals (sodium chloride $Na^+ Cl^-$). When two liquids which are both insoluble in each other are mixed, then they are said to be immiscible. Water and alcohol are similar (H-OH and C_2H_5-OH) and so are miscible. Water and oil are not similar and so are immiscible. When shaken together the oil becomes a suspension in the water (as in salad dressing) but when allowed to settle it will float on top, being less dense than water, as an immiscible layer. Alcohol, being an organic substance should dissolve the oil and allow it to then dissolve in the water. Adding some alcohol to a petrol tank which has had some water leak into it was once used to solve the problem.

CONCLUSIONS:

Summarise these properties of water in a table and give a reason for each of these properties in terms of the water molecule:

PROPERTY	REASON
PART A: Polarity of Water	Ends of the water molecule act as positive and negative ends because of the relative size and nature of the hydrogen and oxygen atoms.
PART B: Surface Tension	Water molecules strongly attract each other because of this polarity and so at the surface the force interaction acts like a skin.
PART C: Cohesion and Adhesion	Water molecules will stick to each other (cohesion) and to other substances (adhesion) depending upon molecular similarity.
PART D: Capillary Action	When forces of adhesion are greater than cohesion, water will move up a tube against gravity as capillary action.
PART E: Heat of Vaporization	Water needs heat (of vaporization) to turn into a gas. This heat is often taken from the surroundings.
PART F: Solubility and miscibility	Water will dissolve or mix with other polar substances but will not mix with those which are not.

RESEARCH: (Optional)

1. How is capillary action used in living things?
The best example is the capillary action (along with transpiration pull) which pulls water up through the stems of plants, including giant trees.

2. What is Paper Chromatography? How is it used as a scientific tool?
Similar to the experiment here, an unknown substance, which may be treated with dyes which are absorbed by specific chemicals, is placed near the bottom of a sheet of special absorbent paper or a column of absorbent powder such as silica gel. The paper or column is then placed in a sealed container with the paper end sitting in a small amount of solvent which moves up the paper or tube and selectively takes the components of the unknown up with it.

See: https://www.youtube.com/watch?v=ZCzgQXGz9Tg
(good, simple experiment)

https://owlcation.com/stem/What-is-Paper-Chromatography-and-How-does-it-Work

https://chemistry.tutorvista.com/analytical-chemistry/uses-of-chromatography.html
(Uses of chromatography)

Experiment 10.2: Mapping the Depths

This is a neat but fictional exercise to simulate how oceanographers use SONAR to plot the depths of the oceans. Prior to this, sea captains needed a considerable amount of line and a lead weight to probe the shallow depths around continents. With the average depth of the ocean being about five kilometres, lead lines were not very practical. Even with modern SONAR and other devices, the ocean depths still have not been fully mapped and hold a lot of secrets. Oceanography is a science which combines the love of the sea with the thrill of scientific discovery.

Students should simply plot a good-sized graph of location against depth by first using reflection time (i.e. the time for the sound waves to go to the sea floor and then return to the detectors below the ship) and the formula to calculate these values. The shape of the graph will be a representation of the shape of the sea floor. In reality, this will be done automatically and will come out of the printer or on the screen as a chart of the sea floor. Echo sounders are common place on many boats.

STATION	DISTANCE TRAVELLED (km)	REFLECTION TIME (seconds)	CALCULATED DEPTH (m)
1. Puerto R.	0	0	0
2.	100	0.3	226
3.	200	4.0	3014
4.	280	5.3	3994
5,	400	6.4	4822
6.	600	6.9	5199
7.	1000	6.6	4973
8.	1400	6.3	4747
9.	1800	5.8	4370
10.	2000	5.1	3843
11.	2100	3.5	1733
12.	2200	3.3	2487
13.	2400	5.8	4370
14.	2600	6.6	4973
15.	3200	6.9	5199
16.	3600	5.9	4446
17.	3800	4.2	3165
18.	3900	0.2	151
19. Orupendo	4000	0	0

For the image reproduced below, a scale of 1 cm = 1000 metres depth was used on the vertical axis and 1 cm. = 200 kilometres was used for distance along the horizontal axis. This gives a vertical exaggeration of 200 and so the angles of the slopes will also be greatly exaggerated.

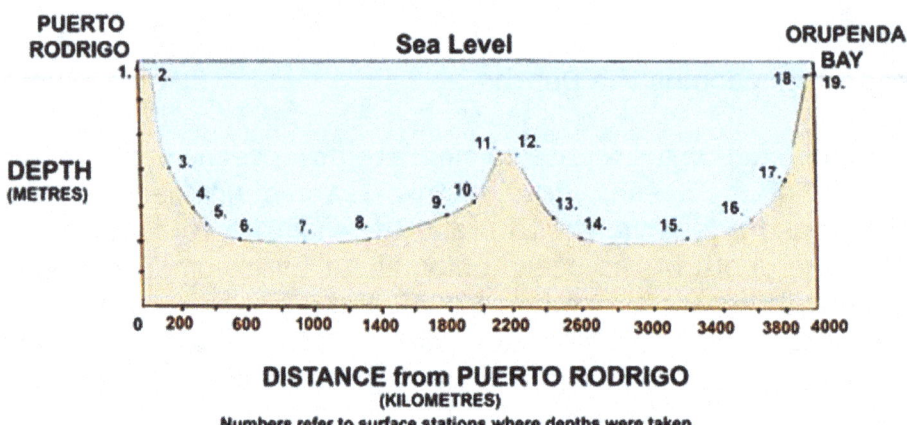

Naturally the vertical exaggeration of this map will be very large considering that the ship has sailed 4000 km and has recorded depths of about 5000 metres. The students should not accept this great exaggeration but try to calculate it (V.E. = about 200). In reality the printouts are also exaggerated because if the same scale were to be used, the sea floor heights would not look very impressive. If a scale of 1 cm = 500 km was used to give a god-sized graph horizontally and the vertical scale was the same (V.E. = 1), then the central ridge would only be about 0.05 mm high on a horizontal axis of 8 cm.

CONCLUSIONS:

1. **What is the main feature seen in this cross-section of the ocean floor? How was it formed?**
   It is probably a mid-ocean ridge formed by the up-welling of basalt which moves out to either side forming the sea floor.

2. **What other oceanographic methods could be used to confirm the nature of this feature?**
   Selective deep-sea drilling and radiometric dating of the cores. The rock should be basalt and the dates would get older moving out from either side of the ridge.

3. **What name is given for the section of this cross-section:**

   a. **just offshore east from Puerto Rodrigo and just offshore west of Orupenda Bay?** - the Continental Shelf.
   b. **between Stations 2 to 3?** - the Continental Slope.
   c. **between Stations 4 to 6?** - the Continental Rise.
   d. **between Stations 6 to 8 and 14 to 15?** - the Abyssal Plain

4. What is the slope of the section between:

   a. Stations 2 to 3? (Hint: Slope = Rise/Run and use trigonometry)?
      Slope between Stations 2 and 3 is about 0.026 (about 1.6^0 using Tan tables).

   b. Stations 4 to 6?
      Slope between Stations 4 to 6 is about 0.0038 (0.21^0). Typically, these slopes are a little bigger, say about 4 and 2 degrees respectively.

5. What is the general depth of this section in 3 (d) above? – about 5000 m.

Experiment 10.3: Some Water Sampling Techniques

This is an introduction to water sampling in the field and, if possible, should be followed up with some field sampling. In a simple approach, students may be able to bring some samples from outside but should be warned in advance about the dangers of sampling water in rivers, creeks and near the ocean. If this is considered risky, then samples could be collected by helpful adult staff members or parents.

PART A: Turbidity

Secchi disks can be made beforehand or as a class activity as described. Some experimentation is needed in attaching the disk/washer to the string. The centre can be pierced and the string inserted through the hole and then tied around a matchstick.

A good substitute for a bucket is a large 500 ml measuring cylinder. Students should be precise in pinching the string at the point when the Secchi disk just disappears and then measuring this length with a ruler.

An additional activity here is to have different groups of students shake the measuring cylinders and at different times afterwards (i.e. at different stages of settling) again measure the Secchi disk depth. They could then combine the results and graph time of settling against Secchi depth to get a rough rate of settling for that sediment.

PART B: pH

This is a measure of the acidity of the water and has very important implications for the biota of natural water which general are neutral (pH = 7). The pH of soils from a previous experiment can be revised here. Universal Indicator solution is useful but the paper version is less messy and more convenient in the field. Refer to the colour chart in the experiment in Chapter 8 or use the link provided.

PART C: Temperature

Use thermometers which have a narrow range of temperature for this experiment and take care not to leave the bucket in the sun for too long.

PART D: Oxygen Content

The preparation of the testing solution could be a problem and is best done before the lesson. In the experiment it can be done as a demonstration using samples from each group. In advance have a sample of water that has been heated to remove the oxygen and then stoppered. Alternatively, it can be done in the classroom to show the bubbles of air coming off the water well before boiling point.

The special chemicals for this experiment should be available from the Chemistry Faculty, but having commercial solutions on hand, especially within a commercial Water Testing kit, is ideal. If this type of activity has not been used before, the commercial kits would be most useful, even if this part of the experiment is done as a demonstration. The idea is to show how water can be sampled for oxygen content, a critical test when it comes to natural waterways receiving fertilizers from agriculture and prone to blue-green algal growths.

https://www.youtube.com/watch?v=9Jk0AmmuLSY
(water testing in the field)

PART E: Phosphorus as Orthophosphate ions (Qualitative Method only)

More useful as a demonstration because of the use of the concentrated nitric acid. Again, if a commercial Water Testing kit is available then it is best used.

PART F: Nitrates - the Brown Ring Test

This is another DIY activity which is best replaced by commercial nitrate testing solutions if available. This is a standard analytical chemistry method which still should work well to show the principle at work within a laboratory setting. It relates also to the last part and the excessive uses of agricultural fertilizers.

PART G: Testing for Bacteria

Some biological skills are needed here! Teachers unfamiliar with this technique should experiment beforehand - it is not difficult.

Agar plates are simply made by the following method(s)

https://teach.genetics.utah.edu/content/microbiology/plates/

http://www.the-odin.com/making-agar-plates-for-bacterial-growth/

https://bitesizebio.com/6938/how-to-make-the-perfect-agar-plate-every-time/

https://www.madaboutscience.com.au/shop/free-experiments/post/grow-bacteria-on-homemade-agar-plates/

Also see videos on this topic:

    https://www.youtube.com/watch?v=yY8STATjZ6U

    https://www.youtube.com/watch?v=YX_b02KYN9g

    https://www.youtube.com/watch?v=ixzhCjLARnU

    https://www.youtube.com/watch?v=0heifCiMbfY
      (using the loop on a prepared plate)

If the classroom does not have a simple incubator then the plates can be placed in a cupboard or closed box in a warm position of the room.

10.4 Other Activities

- Do some simple surface tension experiments such as the coloured milk experiment. Fill a shallow dish (e.g. large petrie dish with milk (mostly water) and, using an eye-dropper, carefully place one drop of different food dyes together at the centre. With a cotton bud soaked in detergent, gently touch the centre of the dyes and watch the colours mix as the surface tension breaks. See:

  https://www.stevespanglerscience.com/lab/experiments/milk-color-explosion/

- Look carefully at a water meniscus. Seen from the side, the water surface will show a curved meniscus bulging upward. A good challenge is to see how much water a student can add to a glass tumbler. It needs to have a completely circular rim so as not to distort the meniscus sides. Once the tumbler has been filled to its brim in the normal manner, an eye-dropper can be used to continue adding water to the 'full' tumbler. A closed-circuit magnifying camera pointing at the tumbler from the side is most dramatic.

- Colour some light oil red or blue with an oil dye or paint. Add to a water colour and shake. A short-lived 'lava lamp'. One needs a dyed wax which is denser than water at ordinary temperatures but less dense when heated. Displaying a real lava lamp will help show this. Turn it on at the start of the lesson as a motivation step.

- Invite an oceanographer or environmental field worker to give a talk.

- Visit the JOIDES (Joint Oceanographic Institutions for Deep Earth Sampling) website for educators and find out about their activities at:

https://joidesresolution.org/for-educators/

https://joidesresolution.org/category/education/

https://joidesresolution.org/

- Watch the author's videos given in the text and discuss the formation of coral reefs, their navigation and some of the recreational uses of coral reefs. Also discuss the difficulties in sailing Tall Ships such as Schooners. See these at:

> Online Video: Dive on the fringing reef at Lifou, Loyalty Islands.
> Go to https://www.youtube.com/watch?v=sAf2Ry5DgbE

and

> Online Video: Sail a schooner to the outer reef of the Great Barrier Reef, Australia
> Go to https://www.youtube.com/watch?v=c9NG-R9046Q

10.5 Answers to Multichoice Questions

Q1.C Q2.D Q3.C Q4.B Q5.A Q6.A Q7.B Q8.D Q9.C Q10.D

10.6 Some Suggestions for the Review Questions

1. **Distinguish between the properties of adhesion and cohesion. Why is it important in regard to the water molecules? Give some examples of both of these forces being used by water.**
   Adhesion is the ability of water molecules to stick to other substances whereas cohesion refers to the ability of water to stick to itself e.g. in capillary action, water will stick to the glass by adhesion but the water molecules of the small meniscus on the surface of the water column will stick to each other by cohesion giving a rounded surface.

2. **Oceanographers often refer to on station and underway activities when it comes to measurement and gathering samples.**

   a. **What do you think are the likely meaning of these two terms; and**
   On Station is when the oceanographic vessel stops at one location and stays 'on station' using a variety of propulsion to keep in one place. Underway is when the vessel is moving and dragging or monitoring.

   b. **List some of the instruments which may be used for each activity.**
   On station instruments usually include corers or grabs or other devices to sample the specific site. Underway, the vessel may use depth SONAR, side-scan radar or dragging nets or sensors.

3. **What factors determine if a coral reef will be formed? Give some locations in the world's oceans where reefs are found.**
   Coral reefs are formed in warm, shallow tropical waters which are clear of turgid sediments.

4. **Northern Scotland and Iceland are much closer to the Arctic than Newfoundland in Canada, yet the environment in Newfoundland is much more hostile than these countries in northern Europe. Explain.**
   There is a warm current (the Gulf Stream) flowing out across the Atlantic, missing the coast of Canada but flowing up past Scotland. See Chapter 14.

5. **Limestone hills and mountain ranges containing marine fossils (corals, brachiopods and the like) are often found well inland and often at a considerable height above sea-level. Use the theory of plate tectonics to explain this observation.**
   These findings occur as the land is pushed upwards by mountain-building processes (orogeny) due to plate collisions at convergent boundaries.

6. **Explain the differences between divergent, convergent and conservative plate boundaries, relating them to the maintenance of crustal material.**
   Divergent boundaries are where plates move apart such as at mid-ocean ridges producing sea-floor spreading as basalt comes up from the crust. At convergent boundaries, one plate may sink below another at a subduction zone removing this plate material. Conservative boundaries are where plate margins move past each other so there is no loss in material.

7. **How does the theory of plate tectonics explain:**

   a. **continental fit** – when a plate splits and moves apart, new oceans and sea-floors form so the present remains of the older continents may be seen as landmasses which could be fitted together if the process was reversed.

   b. **sea-floor spreading** - If a landmass splits into two or more plates, they diverge and new basaltic material comes up at the spreading centre to form new ocean crust.

   c. **the ring of fire** – is a common name for the circle of plate boundaries around the Pacific Ocean where plates are subducting and forming volcanoes on the downward side of the zone thus giving a ring of active volcanoes around the Pacific.

   d. **the existence of *Dicroidium* fossil plant species in India** – this Late Triassic plant is also found in Australia, South Africa, South America and Antarctica because all of this landmass was once joined as the super continent Gondwanaland. These plants represent one genus of the plants which existed in the cool, wet rainforests of that time.

8. Assuming that climate change will mean an increase in the global temperature, what are the probable effects that this will have on:

   a. **the land** - generally will become hotter and drier in inland districts and probably wetter along tropical coasts.

   b. **the oceans** - uniformly becoming warmer with a higher carbon dioxide content and so more acid.

9. Evidence suggests that the breakup of Pangaea was only the most recent of several older episodes. Suggest current locations in the world which might represent new plate formation or disintegration (Hint: use the internet and look for information on the Wilson Cycle and rift valleys).
   Some ancient relic structures in and around old craton regions of the world such as near the Urals in Russia and parts of central China suggest older plate boundaries now accreted together. Many of the older Palaeozoic rock regions of the world suggest multiple accretions. See:

   https://theconversation.com/plate-tectonics-new-findings-fill-out-the-50-year-old-theory-that-explains-earths-landmasses-55424

   https://pubs.usgs.gov/gip/dynamic/Pangaea.html

10. Use the Internet to research how modern oceanography is used by the world's maritime nations and how does this relate to the use of economic resources?
    There are a great many functions carried out my oceanographers apart from scientific research and monitoring weather and tsunamis. Most Navies have oceanographic branches to chart channels, sea floor regions and layers of salinity to hide or detect submarines. Many mining companies also have oceanographers to look for valuable minerals on the sea floor such as manganese nodules and for possible sub-surface oil traps for off-shore oil and gas. See:

    https://www.nationalgeographic.org/encyclopedia/oceanography/

    https://www.nsf.gov/geo/oce/whatis/tools.jsp

    https://skymapglobal.com/oceanography-remote-sensing/

10.7 Reading List

Able, K.W., et.al, (1987), Sidescan-sonar as a tool for detection of demersal fish habitats. *Fishing Bulletin*, v. 85, n. 4.

Ahrens, C. Donald, (2007). Meteorology today: an introduction to weather, climate, and the environment. Cengage Learning. pp. 296. ISBN 978-0-495-01162-0.

Australian Bureau of Meteorology (BOM): http://www.bom.gov.au/australia/

Daly, R. Terik; Schultz, Peter H. (25 April 2018). "The delivery of water by impacts from planetary accretion to present". Science Advances. 4 (4). doi:10.1126/sciadv.aar2632.

Dodds, D. (2001). *Modern Seamanship: a comprehensive ready-reference guide for all recreational boaters*. Guilford, Conn.: The Lyons Press. ISBN 9781585745289.

Fish, J.P. and H.A. Carr, (1991). *Sound Underwater Images, A Guide to the Generation and Interpretation of Sidescan Sonar Data*. Second edition. Orleans, MA: Lower cape Publishing.

Gnanadesikan, A., Slater, A., R. D., Swathi, P. S.and Vallis, G. K. (2005). The Energetics of Ocean Heat Transport. *Journal of Climate* 18 (14): 2604-16. Bibcode: 2005JCli...18.2604G. doi:10.1175/JCLI3436.1.

Hallis, Lynda J. et al. (2015). *Evidence for primordial water in Earth's deep mantle.*
Science 13 Nov 2015: Vol. 350, Issue 6262, pp. 795-797.DOI: 10.1126

Heinemann, B. and the Open University. (1998). *Ocean Circulation*. Oxford University Press.

Institute of Astronomy University of Hawaii
http://www.ifa.hawaii.edu/info/press-releases/water-origins/

Lurton, X. (2002). *An Introduction to Underwater Acoustics, Principles and Applications*. Springer in association with Praxis Publishing, pp 680. ISBN 978-3-540-78480-7.

Matsui, Takafumi & Abe, Yutaka. (1986). Evolution of an impact-induced atmosphere and magma ocean on the accreting Earth. Nature. 319. 303-305. 10.1038/319303a0.

National Geographic. (2009). *North Magnetic Pole Moving East Due to Core Flux*. December 24.

National Aeronautics and Space Administration (NASA): https://www.nasa.gov/

National Oceanic and Atmospheric Administration (NOAA): http://www.noaa.gov/

Pinet, P. R. (1996). *Invitation to Oceanography*. Eagan, MI: West Publishing Company. ISBN 978-0-314-06339-7.

Rice, A.L. (1999). *Understanding the Oceans: Marine Science in the Wake of HMS Challenger*. Routledge. pp. 27-48. ISBN 978-1-85728-705-9.

Sarafian, Adam R.; Nielsen, Sune G.; Marschall, Horst R.; McCubbin, Francis M.; Monteleone, Brian D. (2014-10-31). *Early accretion of water in the inner solar system from a carbonaceous chondrite-like source*. Science. 346 (6209): 623-626. ISSN 0036-8075. PMID 25359971.

Scott, P.T. (2018). Through Sea and Sky. 2nd Edition. Brisbane: Felix Publication. ISBN: 978-0-9946432-2-3

Shiklomanov, Igor. (1993). *World fresh water resources*. In Gleick, P.H. (editor). Water in Crisis: A Guide to the World's Fresh Water Resources. (Oxford University Press, New York)

Sverdrup, K. A., Duxbury, A. C. & Duxbury, A. B. (2006). *Fundamentals of Oceanography*, McGraw-Hill, ISBN 0-07-282678-9

Urick, R.J. (1983). *Principles of Underwater Sound*. Peninsula Pub., 423 pp.

Thurman, Harold and Alan Trujillo. *Introductory Oceanography*. 2004. p151-152. Prentice Hall; 5th edition (1996). ASIN: B000OIPPTO

Chapter 11:   The Atmosphere - The Air Above

11.1 Theme

The atmosphere is the blanket of air which covers the planet and consists mainly of nitrogen with 21% oxygen, water vapour, carbon dioxide and other gases. The study of the atmosphere and all of the material and energy processes which cause the world's weather is called meteorology.

11.2 Rationale

An understanding of how the atmosphere and the world's weather works is vital to most Human activity as well as to an understanding of most of the processes of the other spheres. There is a constant interaction between the land, the sea and the biota of the planet and its atmosphere.

11.3 Notes on the Practical Work (see PRACTICAL MANUAL 1.)

Experiment 11.1: Relative Humidity
The amount of water vapour in the air is called its relative humidity. Apart from being a major part of the world's Water Cycle, it also is a major part of Human weather records and forecasts. It is part of the comfort effect of the weather and also has an influence on those commercial and industrial processes which either need a moist climate or do not. Moisture is a disadvantage on the production, storage and use of many materials produced commercially. The advantage of this experiment is that it can be performed anywhere as a humidity-measuring system provided that one has two thermometers; one with a damp cloth hanging in water, one dry and a copy of the tables given in this book.

This activity should be done in a room which is not air-conditioned or in the shade outside. Direct sunlight would give a false reading in the thermometers as well as drying out the wet bulb.

CONCLUSIONS:

1) What is the relative humidity:

    (a) at the start of the lesson?
    (b) near the end of the lesson?

2) If there was any change, explain why this may have occurred?

3) If appropriate, was there any difference in humidity between an air-conditioned room and outside? If so, why?

These three questions are best answered by experimentation but it may be expected that in a room that is not air-conditioned but closed up that the relative humidity (and carbon dioxide levels) will rise slightly due to the exhalation of those inside. Air conditioners also dry out the air in a room and so it would be expected that once the AC is turned on that the relative humidity will drop. Try it!

4) Explain why it is better to do the washing on a day with low relative humidity?

If the air contains less moisture then there is more space for the water molecules on the wet washing to move into the air by evaporation, especially if the wet clothing is heated by the Sun and/or blown by the wind. Both of which will allow water molecules to vaporize more freely. If there is a high relative humidity, the air is already saturated with water and this prevents water molecules from leaving the wet surface of the washing.

RESEARCH: (Optional)

Use the Internet and class discussion to consider what household or industrial processes which require low relative humidity.

Apart from washing, there is also any process which requires water evaporation such as convection air conditioning, storage of any product that would attract mould and fungi, processes such as house painting, inhibiting dust mites and other allergens.

See also:

http://www.level.org.nz/passive-design/controlling-indoor-air-quality/humidity-and-condensation/

https://www.sensitivechoice.com/indoor-humidity/

http://www.ahrinet.org/App_Content/ahri/files/Humidity_Occupants_Presentation.pdf

http://energysmartohio.com/how_it_works/why-humidity-matters/

Experiment 11.2: The Aneroid Barometer

This is another simple DIY experiment but it shows the concept of the barometer and fluctuations in air pressure with time. There should be no problems except with gluing the drinking straw onto the stretched balloon. If a standard specimen large jar is used then the normal-sized balloons cut open should stretch across the mouth nicely. It should also be secured with a thick rubber band or some tape around the side of the jar's mouth. The other end of the drinking straw can be sliced to give a flat point as an indicator. The set up should be in a room or other place where the air pressure would change. Sealed, air-conditioned rooms might not be useful. Also, it might be just a coincidence that over a few days that the air pressure outside remains constant. Students could watch the nightly news or use an App. to record the local air pressure for homework.

CONCLUSIONS:

1. Did the relative air pressure change over several days? How?
This will depend upon the local weather changes. It would also be useful to calibrate the instrument by comparing changes observed with the actual changes in air pressure outside.

2. What would be the errors in this experiment or the disadvantages of such a device for measuring air pressure?
As a DIY instrument, measurement is not reliable but the instrument error relates to the ruler scale and how it is aligned. The device will only show changes in air pressure and then only slight as any change is relative to the original air inside the jar.

3. How could this crude instrument be calibrated?
As suggested, it could be roughly calibrated using news weather reports or the local pressure given by the weather services and this pressure marked on the rule. At least two widely different pressures would be needed for this.

4. What would be the advantage of having a jar with less air pressure inside (as in the box of a real Aneroid Barometer)?
With little or no air inside the box, the differences between it and the outside air pressure would allow for a greater variation in changes of air pressure.

RESEARCH: (Optional)

Use the Internet to find out more about barometers, comparing the usefulness of Aneroid Barometers against Fortin Barometers.
Aneroid barometers are more robust, compact and portable. Fortin barometers have few moving parts and so tend to be more reliable and more accurate. It is common practice to lightly tap the glass of the Aneroid barometer to loosen any levers or springs which may have become stuck due to the humidity. On the other hand, imagine carrying a large column of mercury in a delicate glass tube up mountain ranges and into jungles like some of the early researchers.

See:

http://www.bom.gov.au/australia/index.shtml

http://www.bom.gov.au/watl/pressure/index.shtml
(Bureau of Meteorology - for Australia)

https://w1.weather.gov/data/obhistory/KBOI.html
(NOAA – United States)

http://resource.npl.co.uk/pressure/pressure.html
(for the UK)

Experiment 11.3: Weather Monitoring

This is a good exercise in keeping records such as weather data over time. It could be done using a paper wall chart, book record as per this experiment or electronically. It might involve a project setting up a regular weather station in a secured area under cover and in the shade. Electronic weather stations are common and can be purchased from local hobby stores. The ideal situation is a designated Stevenson Screen with a set of instruments (maximum-minimum thermometer, aneroid barometer, wet/dry bulb hygrometer) and nearby wind and rain gauges.

CONCLUSIONS:

1. Has the weather changed over the week? How (give details)
2. Was the weather in the next week as predicted? Comment.

This will depend upon recorded results and the local weather. The main idea here is making records and predictability.

3. List all of the problems or errors which could occur with such short-term weather measurement.

There are local influences on weather instrument such as local air conditioning, exhausts and too many people nearby. Of course, the weather may not vary too much over such a short time.

RESEARCH (Optional):

1. Where is the place where the weather is measured and recorded locally?

In cities, the weather is recorded at a number of government and private sites. Usually at airports and government weather agency offices. In country areas the airport and post office are the usual places for recording weather data.

See:
http://www.bom.gov.au/climate/data/index.shtml?bookmark=200&view=map

2. How is weather data gathered on a wider scale?

Weather details are also recorded remotely from buoys at sea, remote stations on islands and in the interior as well as by ships at sea.

See:
http://www.bom.gov.au/climate/cdo/about/map-search-guide.shtml

https://www.noaa.gov/resource-collections/climate-monitoring

https://www.weather.gov/
(for the United States)

3. How are weather predictions made?

Meteorologists have measured the weather for many years and have been able to notice trends in local and national weather patterns. Using computer algorithms, they are able to make predictions based upon these trends related to current measurements.

See:
http://www.weatherquestions.com/How_are_weather_forecasts_made.htm

https://www.metoffice.gov.uk/learning/making-a-forecast

http://www.minitab.com/en-us/Published-Articles/Weather-Forecasts--Just-How-Reliable-Are-They-/

Of some interest is the ancient practice of noting animal indications of near future weather. For example, ants seem to climb upwards just before rain and birds flock together and fly off in front of approaching storms. Many local farmers have their own lore about predicting weather and field scientists could find such local knowledge when going well away from shelter into weather-hostile zones. For example, parts of mountainous south-eastern Australia and Tasmania are subject to sudden weather changes, especially in summer when a wind change can drop the temperature from hot to below zero.

See:
https://www.outdooraustralia.com/articles/Predict-the-weather-without-a-forecast-05041

http://www.abc.net.au/science/articles/2010/05/18/2902595.htm

http://www.howtoplaza.com/how-to-predict-weather-by-observing-animal-behavior

11.4 Other Activities

- Make a saturated solution of cobalt chloride and soak filter paper in the solution in a flat dish. Hang the wet papers up to dry and they will change to a blue colour. Students can take these papers home and hang them in a sheltered, shady spot away from other sources of humidity such as in the kitchen – but remind them that the chemical is not to be tasted. They are relatively good indicators of humidity as they will turn purple to red as the humidity increases and back to blue when it is low.

- Visit a local weather-recording site by prior arrangement such as a meteorological office or post office or have a meteorologist visit.

- Make fog by filling up a wide-mouth jar completely with hot water for about a minute. Pour out almost all the water, leaving about 2-3 cm in the jar. Put a wire strainer over the top of the jar and place a few (3 or 4) ice cubes in the strainer. As the cold air from the ice cubes collides with the warm, moist air in the bottle, the water will condense and fog will form.

- Make a simple water thermometer by pushing a length (say 20 cm) of capillary tubing through a drilled cork or stopper and inserting it into a round flask which has about one third full of red coloured water. Place a paper scale on the side of the upper part of the tubing and note daily changes.

- Also see this site which has many simple experiments and links about weather:

    https://www.teachervision.com/all-kinds-weather

11.5 Answers to Multichoice Questions

Q1.C   Q2.A   Q3.D   Q4.A   Q5.C   Q6.B   Q7.B   Q8.A   Q9.D

11.6 Some Suggestions for the Review Questions

1. **List some of the equipment and methods of meteorologists. Discuss how such a career would adventurous in some cases.**
   Many of the instruments have been mentioned but in addition equipment will also include drones and radiosondes with balloons which carry remote-sensing equipment and radar tracking reflectors. Because much of the world's weather is often generated in remote places such as oceans, alpine regions and places like Greenland and Antarctica, meteorologists often find themselves in remote places with extremes of weather.

2. **Much of the world's weather data comes from remote sensing devices. Give an example of such a device and explain why meteorologists must rely upon such data.**
   Many weather stations are set up on remote islands and in hostile places such as in Antarctica. The data from these stations are relayed back to a meteorological station via satellite link. These are set up for continuous recording in places where it would be unsatisfactory for Human habitation during all seasons.

   See:
   https://rmets.onlinelibrary.wiley.com/doi/full/10.1002/met.288

   https://www.slideshare.net/rakhighosh/remote-sensing-and-weather-forecasting

3. **Compare and contrast the Fortin Barometer and the Aneroid Barometer.**
   Both instruments can be used to measure air pressure in reliable units to note changes in pressure. The Aneroid is robust, compact and portable whereas the Fortin barometer is larger, bulky and more delicate. It is, however more accurate and has no moving parts to become stuck.

4. What is meant by each of the following terms:

   a. **hectoPascals** - (symbol hPa) is the SI unit used in the measurement of air pressure and is approximately equal to one millibar. Normal air pressure is 1013 millibars.

   b. **relative humidity** - the amount of water vapour present in air expressed as a percentage of the amount needed for saturation at the same temperature.

   c. **Isobars** – lines joining places of equal air pressure.

   d. **noctilucent clouds** – these are the highest clouds in Earth's atmosphere, located in the mesosphere at altitudes of around 76 to 85 km and are visible at night after twilight when illuminated by sunlight from below the horizon.

   e. **anacoustic zone** - is the region of the atmosphere above about 160 kilometers where the air density is so low that air molecules are not close enough to support transmission of sound waves within the hearing range.

5. **What would be the relative humidity if a wet-and-dry bulb hygrometer showed temperatures of 17^{0C} and 25^{0C} respectively? Would this be a good day for washing clothes? Explain.**
   Relative Humidity would be 44% and a fair day for washing as there would be a reasonable amount of evaporation.

6. **The Planetary Boundary Layer is found at various heights above the Earth's surface. What factors influence its altitude?**
   It is mostly affected by shape of the land and its height as well as the amount of heat radiation coming off the land and sea.

7. **The Earth has several protective shields from harmful things from outer space. What are these and how do they protect living things on the Earth's surface?**
   The magnetosphere due to the magnetic field of the Earth protects the planet from harmful radiation particles due to the solar wind and the Ozone layer also protects against harmful UV rays.

8. **What are CFC's? What were they originally used for and why were they banned?**
   Chlorofluorocarbons are complex non-toxic organic compounds of carbon and hydrogen with linked halogens atoms such as chlorine and fluorine. There were mostly used as refrigerants, as propellants in aerosol cans, in packaging and as solvents. It was found that when they went into the upper atmosphere they reacted with ozone and depleted the protective ozone layer.

9. **Research the internet to find out the CURRENT state of the ozone layer, especially reports about the hole in this layer above Antarctica.**
   The state of the ozone layer is measure in Dobson Units and is currently about 200-300 units as mid-range. The first link gives the current situation and also details of thus unit:

   https://ozonewatch.gsfc.nasa.gov/

   https://ourworldindata.org/ozone-layer

   https://theconversation.com/is-earths-ozone-layer-still-at-risk-5-questions-answered-91470

   http://www.cpc.ncep.noaa.gov/products/stratosphere/sbuv2to/ozone_hole.shtml

10. **Research the internet to find out about the atmospheres of Venus and Mars. Why did these planets not evolve an oxygen atmosphere?**
    The atmosphere of Mars is about 100 times thinner than Earth because it is a very small sized planet. The thin atmosphere is mostly carbon dioxide (a relatively heavy gas) but there appears to be some water ice in parts of the surface. Venus on the other hand has a very thick atmosphere of mostly carbon dioxide and is about 90 times the density of that of Earth. The planet is about the same size as Earth but is much closer to the Sun giving high temperatures due to radiation but also because of the extreme greenhouse effects of the atmosphere.

    See:
    http://www.tp.umu.se/space/Proj_10/Amir_A-10.pdf

11.7 Reading List

Ahrens, C. Donald, (2007). *Meteorology today: an introduction to weather, climate, and the environment.* Cengage Learning. pp. 296. ISBN 978-0-495-01162-0.

Annenberg Foundation (2017). The Habitable Planet: Unit 2: Atmosphere (online learning)
https://www.learner.org/courses/envsci/unit/text.php?unit=2&secNum=0

Australian Bureau of Meteorology (BOM): http://www.bom.gov.au/australia/

Barry, R. G.; Chorley, R. J. (1971). *Atmosphere, Weather and Climate.* London: Methuen & Co Ltd.

European Space Agency (2018). Earth's Atmosphere: New Results from the International Space Station.

https://www.esa.int/Our_Activities/Human_Spaceflight/Research/Earth_s_atmosphere_new_results_from_the_International_Space_Station

Lutgens, Frederick K. and Edward J. Tarbuck (1995) *The Atmosphere*, Prentice Hall, 6th ed., pp. 14-17, ISBN 0-13-350612-6

Meteorological Service, Singapore (2018). Earth's Atmosphere. http://www.weather.gov.sg/learn_atmosphere/ (scroll down to the bottom for different topics and data)

NASA (2018). *First Direct Proof of Ozone Hole Recovery Due to Chemicals Ban* https://www.nasa.gov/feature/goddard/2018/nasa-study-first-direct-proof-of-ozone-hole-recovery-due-to-chemicals-ban

US National Oceanographic and Atmospheric Administration. https://www.noaa.gov/

Sanchez-Lavega, Agustin (2010). *An Introduction to Planetary Atmospheres*. Taylor & Francis. ISBN 978-1-4200-6732-3.

Scott, P.T. (2018). Through Sea and Sky. 2nd Edition. Brisbane: Felix Publication. ISBN: 978-0-9946432-2-3

Timothy W. Lyons, Christopher T. Reinhard & Noah J. Planavsky (2014). *Atmospheric oxygenation three billion years ago"*. Nature. 506 (7488): 307-15. Bibcode:2014Natur.506..307L. doi:10.1038/nature13068. PMID 24553238.

Weather Online (2018). Earth's Atmosphere.
https://www.weatheronline.co.uk/reports/wxfacts/The-Earths-Atmosphere.htm

Zahnle, K.; Schaefer, L.; Fegley, B. (2010). *Earth's Earliest Atmospheres*. In Cold Spring Harbor Perspectives in Biology. **2** (10): a004895.

Chapter 12: The Biosphere – Life on Earth

12.1 Theme

Life exists on planet Earth because conditions were just right for it to happen. The Earth was the right size and distance from the Sun to retain a thick atmosphere and allow for liquid water to exist on the planet. The Earth's magnetic field and atmosphere also protected the surface from harmful charged particles and ultra-violet rays. Life exists in a great variety of forms in many habitats of ecosystems around the planet and whilst many have adapted to moderate climates, extreme forms can be found in high and low temperature conditions deriving nutrients from a variety of sources.

12.2 Rationale

The basic necessities for life come from all of the spheres of the Earth. To understand the requirements of living things is to know how they have developed over time and how they relate to each other and to the abiotic surroundings.

12.3 Notes on the Practical Work (see PRACTICAL MANUAL 1.)

Experiment 12.1: Introduction to the Biological Microscope
There are many tools which need to be mastered by the practicing Earth and Environmental scientist. Within the biotic realm, the most important instrument is the biological microscope. These come in many forms from the simple student monocular microscope, through to the binocular research microscope to the advanced electron microscope which uses electron beams rather than light to scan the specimen. This experiment is an introduction to the simple student microscope and show how living things can be safely examined. It is very important that the students know the parts of the biological microscope, how to carry them and how to use them carefully, especially how to focus the microscope by <u>focusing upwards</u>.

In modern laboratories microscopes are often set up in temperature-controlled rooms but in all cases, they should be kept out of the sun and illuminated by natural light or special lamps – with good use of the condensing lens and mirror. The field of view is important because students may have to draw representations of the specimens or their parts to scale. In modern laboratories, photomicroscopy uses video cameras connected to computers or use other digital cameras. It is a relatively simple task to hook up a small computer video camera to the eyepiece of a good microscope using cardboard and gaffer tape to connect the two. Attachment lenses can also be purchased for mobile phones so that these can be used as powerful hand lenses/microscopes in the field which can then photograph specimens as required.

CONCLUSIONS:

1. What happens to the light intensity and clarity of image as the magnification is increased?
The light intensity decreases as higher magnification is used because of the nature and number of the lenses in the objective. For most simple activities, low power is the best to use (say no more than x100).

2. What is the diameter of the human hair examined?
This will vary depending upon the individual and is a good comparison across the class noting that thick or thin hair has no connotations of superiority. Hair will vary in diameter from 17 to 181 microns (μm - millionths of a metre).

3. Compare the movement of the glass slide with the observed movement of its image in the field of view.
Students will note that it is laterally inverted due to the way the light passes through the lenses. This may be strange at first but they soon learn how to move the slide about in opposite directions.

4. Is there any difference with hair of a different colour?
This is another good class discussion. In general, black hair is thicker than light-coloured hair but then thickness also increases with age and some other factors. It would be interesting have the students draw a column graph for thickness against colour (say Black, brown, blonde and red) assuming that they all are about the same age and in good health.

Experiment 12.2: Examination of a Plant Cell - Onion Epidermal Cells

This is a traditional biological experiment to look at the shape and components, or organelles of a typical plant cell. Onions are used because they can be peeled off into very thin layers and are convenient, but take care of onion juice and the eyes. The secret in this experiment is in the peeling of a single epidermal layer. Tincture of iodine is relatively harmless but it will stain skin, books etc. It is important that the cover slip be put on its edge and quickly dropped onto the stained specimen, otherwise bubbles will be created. These look like huge truck tires through the lens. Begin with low power and perhaps go to medium but always remember to be careful when changing lenses that they do not grind across the slide. Also remember to watch how the objective lens is lowered to the top of the cover slip and then focused <u>upwards</u>.

CONCLUSIONS:

1. Why was iodine used?
Iodine is a convenient stain which is non-toxic and will stain the nucleus of the cell.

2. What distinguishes the nucleus from any other organelle observed?
It will be seen as a round body in the middle part of the cell. Internal features such as chromosomes may be seen in some very good conditions. A commercially prepared slide or photomicrograph of this is good for a demonstration.

3. Comment on the general shape of the cells and how they are packed together. What factors determine the size and shapes of these cells?
Cell packing usually determines the shape but in plant cells with a semi-rigid cell wall, they will have a polygonal shape. The size of each cell is determined by its age and the number of cells around it.

See also:
https://en.wikibooks.org/wiki/School_Science/How_to_prepare_an_onion_cell_slide

https://www.youtube.com/watch?v=dxv4M4HHUgs

https://www.youtube.com/watch?v=zXAgbSzEADQ

Experiment 12.3: Examination of an Animal Cell - Human Cheek Epidermal Cells
Warning! In some schools, the administration does not allow Human experimentation for emotional and health and hygiene reasons. This experiment is also a traditional one and there should be no health problems provided that the students retain their own swabs and that these and the glass slides are disposed of in an appropriate container and manner. This method is the preliminary technique used in identifying genetic features but uses better microscopes and methods.

Methylene Blue is a strong stain, especially on student's hands, books and clothing. Gloves would be useful to prevent this but can be cumbersome. It is better for the students to learn how to be careful and considerate of others.

CONCLUSIONS:

1. Why was 1% saline solution used to cover the glass slide?
It is about the same as Human body fluids so this ensures a good surrounding medium. Distilled water used in the preparation of the stain will burst the cells by osmotic pressure so a little saline solution will prevent that.

2. Why was methylene blue used?
It is a good water-based stain which show up the nucleus of the cell.

3. What distinguishes the nucleus from any other organelle observed?
It will be round, near the centre of a full cell and stained blue. In older cells, both plant and animal there may be some vacuoles of water which will push the nucleus away from the centre.

4. Comment on the general shape of the cells and how they are positioned in the field of view.
In animal cells with no cell wall, their shape depends on the packing of other cells around them.

5. Compare and contrast the size, shape and features seen within these cheek cells from those of the plant cells seen in the previous experiment.
The comparison will concern the shape due to the cell wall of plants but the nucleus and perhaps some other structures such as the granular cytoplasm may also be seen.

See also:
https://www.youtube.com/watch?v=C6-Nat8fwZw

https://www.microscopemaster.com/cheek-cells-microscope.html

Experiment 12.4: Classification of Organisms
There have been many different methods and models for classifying organisms and the systems keep changing as new organisms are found or old organisms are re-matched with others. The classification model used in the textbook appears to be the latest and most useful. It has been detailed with an emphasis on ancient lifeforms but is still appropriate for modern life.

Living specimens are often difficult to observe in a classroom because of their activity so this is why preserved specimens or photos have been used. Some students may be squeamish with preserved dead specimens, especially if the older method of preservation using of toxic formalin was used. The best system is a class set of dried and embedded specimens. It is important to develop a sense of order and the need for classification along with flexibility in approach in this experiment by keeping ideas and systems in a simple context.

CONCLUSIONS:

1. What were the main distinguishing features which were used to differentiate between the classification types of:

    (a) Kingdoms? – some common feature of all members such as how they obtain nutrients.

    (b) within each Kingdom? – more variety here but reproduction and body features are the main criteria.

2. Comment on any difficulties found in this activity.
Often some of the main distinguishing features have been obscured or lost during preservation. Good photographs are ideal as a support. Moreover, dead organisms do not show some of the movements nor habits of the live specimen.

3. Why are classification systems often changed? Is the one used in the text the ultimately best system? If not, why not?

Classification systems often change as new specimen are discovered or if taxonomists change some of the criteria for classification to match convenience. In addition, scientists often used slightly different classifications which are more suited to their field of study. The one used in the textbook comes from the author's perspective as a palaeontologist and this could change because of a new perspective or if new species are described. The world is still a very wide environment with large areas yet to be researched such as the Amazon jungle where new plants, insects and other animals are constantly being discovered. The seas are also another source of unknowns.

Experiment 12.5: Examination of Some Common Fossils

This should be a major activity as it introduces palaeontology as a science, uses some of its laboratory methods and teaches students important skills of observation, description, sketching and research.

The use of sketching in this modern electronic age should not be overlooked as it provides students with a fine motor skill/art and enables better observation and recording of detail. Some students may wish to photograph their specimens (this depends on school policy about the use of cell phones in the classroom). The author has often allowed his students to photograph specimens but only as an additional aid to the sketches. Students who are using electronic notebooks in the classroom will have to complete the sketches by hand in such a clear manner (all students might enhance their sketches by going over the outline and main detail in heavy pencil or black ink) that they can then be photographed on site or scanned electronically at home for inclusion in their reports (which may/may not be submitted electronically, depending upon the teacher's policy)

A short note on sketching skills follows. These notes may be photocopied or cut and pasted to upload to students if required or simply used on a smartboard, large computer screen or data projector directly from this book.

SKETCHING SKILLS

Sketching is a scientific SKILL which usually has to be taught. The following steps should be used for each of the specimens.

1. Carefully examine the specimen for major detail (non-rock features such as lines, curves, shapes, colour and sections resembling known life-form features e.g. eyes, legs, plant veins etc.). Consider the following example (*Lepidodendron, a tree-fern*):

e.g. scale = **x1**

2. Draw the <u>outline</u> of the entire rock to an appropriate scale so that it takes up about one-third of a page. It may be scaled up (e.g. X 2 such that it is twice as long and wide as the original) and edges of the rock drawn in to give a 3D effect.

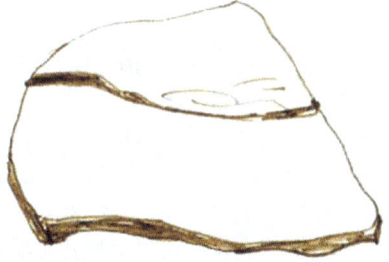

3. Details of the fossil can then be drawn in using faint construction lines and geometrical shapes (circles, squares, rhombuses etc.) or lines and curves. Look for the detail but do not draw everything if the fossil is extensive, draw only the outline and a small representative area.

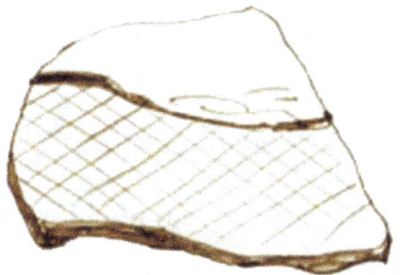

4. Complete the details (as appropriate) and shade in sections for effect. The main features and outline of the rock may be redone in heavier pencil or black ink (if good!).

5. Add some labels (PRINT IN UPPER CASE in ink) of main parts identified and any useful information. Give the name of the organism in *Italics*. These are obtained from texts.

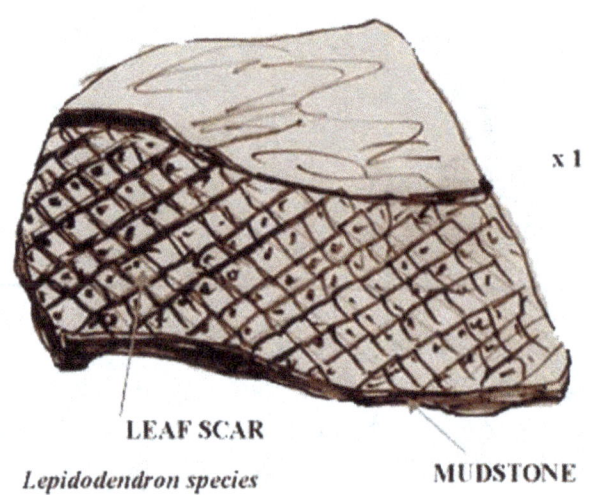

LEAF SCAR

Lepidodendron species **MUDSTONE**

The activity requires a class set of identical specimens (say about 8-10 each set) for each group in the class (10?). Each set should be boxed (say in small identical cardboard or plastic containers such as food containers etc.) and numbered (e.g. box 1, 2, 3 etc.) and each specimen lettered (e.g. 1A is specimen A in box 1 etc. or some system) so that identical specimens have the same letter in each numbered box. This can be done using a circle of white (or other colour) upon which the code (e.g. 1A, 2B, etc.) is written in black paint or permanent ink.

There should be a good variety of specimen types - old (say Permian) plants, young plants (say Triassic), a good variety of invertebrates (which could include trilobite, graptolites, corals, bryozoans, echinoderms, brachiopods, molluscs. etc.) and if possible, a vertebrate or part thereof. Any or all of these may be replicas painted and lacquered to resemble the original.

Replicas can be made from the one good specimen (use those with good 3D features not a simple flat surface) by:

1. Placing the original on a piece of cardboard.

2. Liberally coating it with cooking oil (no bubbles) using a fine brush.

3. Pouring or dabbing on a layer of latex (can be purchased from craft or hobby stores as moulding latex or for small scale application, use a tube of carpet glue latex).

4. Cover with strips of tough fibrous cloth like that used in cleaning, Chux NOT tissue paper), painting it on with more latex. This is done until the layers seem thick enough.

5. When fully dry (may take overnight), carefully peel off the rubber mould and allow to dry inside. More layers may be added to the outside if required. Lightly powder and store in a sealed container so that moulds do not touch each other).

Casts can be made from these moulds using plaster of paris which can then be painted in appropriate colours using water-based paints, labelled and then lacquered with a thin coating of spray clear enamel. A set of 10 replicas can be made very quickly, especially if the class is used as a palaeontology workshop activity.

Originals and replicas are sold on the internet at often inflated prices. Schools wishing to purchase fossils should use their Earth Science supplier or take great care in shopping online. Specimens need not be rare nor over expensive but should clearly show the main features of the species and be of a reasonable size (hand specimen). If the school has sufficient funds then some bigger specimens may be purchased for display (don't pay more than $10 unless an excellent find).

Some advice and sales can be found at:

http://www.fossils-facts-and-finds.com/buy_fossils.html (good advice and site)

http://www.ebay.com.au/gds/Buying-Fossils-Law-Ethics-Forgeries-/10000000001926697/g.html (search eBay under separate type e.g. trilobites – there are some good buys and a lot of expensive items to ignore!)

https://english.fossiel.net/information/article.php?id=79&/Buying%20Fossils

https://www.ukge.com/en-au/Fossils-for-Sale/Popular-Fossils__c-p-0-0-15-291.aspx (sets at fair prices)

http://livingfossil.com.au/fossils-sydney-living-fossil-crystals-fossils/fossils-under-50-sydney-living-fossil-crystals-fossils.html

http://www.fossilmall.com/fossilpurchase.htm

In some cities, local fossil clubs or universities may allow a staff member to make quick moulds of good specimens (say a set of trilobites) in modelling clay (plasticine) which can be softened in the hand and then hardened in air or even a refrigerator. Then, a cardboard rim about 1.0 cm. high can be circled around the mould and plaster added. This should give a good replica but may destroy the mould. New moulds can then be made using latex.

CONCLUSIONS:

1. **Compare and contrast the use of originals and replicas in this study if appropriate. (Are replicas useful? Why?)**
   Replicas vary greatly in their accuracy depending upon how roughly they are made. Originals are always better as they usually show good detail. Replicas are useful when the originals are hard to get, expensive and are easily broken.

2. **Why must the scale be given for each specimen?**
   Practical Reports are another communication tool in Science. A reader needs to know the original size of the specimen.

3. Give uses for such a study of fossil specimens.
Here they are being used to reinforce theory notes and pictures and give the student the chance to obtain real detail of an ancient organism which is often rare to find.

4. What are some problems with such a study? Could it be improved?
The problems associated with this study are to do with detail. Specimens must have good detail and show what they intend to show. Poor specimens or badly-made replicas are not good. The sketching skill of the student may also lower the standard of their report. In reality, high-resolution and 3D photography is often used to record the details of a rare specimen.

5. If this were a major work, what other research activities would be necessary to fully give more precise details of exact species, age and environment?
The specimen may be subjected to a more detailed study using high quality stereo microscopes (even electron microscopes for smaller specimens) and the rock containing the specimen (or even part of the specimen) may be radio-metrically dated to determine its absolute age.

1.4 Other Activities

- Examine the leaves of a pondweed algae such as *spirogyra* or *elodea*, as wet mounts to look for chloroplasts containing chlorophyll.

- Use bubbles to see the effect of packing on flexible spheres. This is done by wetting the surface of a smooth tray with a bubble solution.
  See:

https://www.questacon.edu.au/outreach/programs/science-circus/videos/bubble-mix-recipe

  Use a drinking straw to blow bubbles together onto the tray and note the shape of bubbles in the centre. If the end of the straw is wet with bubble mixture, it can be poked through one large bubble and another bubble can be blown inside i.e. a vacuole?

- Look at commercially-prepared slides of various small animals such as insects, blood cells, bacteria and other organisms.

- Invite a research medical professional in to talk about the use of microscopes in research and/or the use of cheek swabs to test DNA.

- Plan an excursion to a nearby fossil bed. This will take some care as conservation of the beds is very important. A formal excursion with precise objectives needs to be designed and a risk assessment to the students (and also the environment) needs to be done by the teacher who should visit the area well before the event. A minimal number of students should be encouraged to only take minimal numbers of specimens and walk around the site carefully. Perhaps only 3-5 small specimens would be needed by each student and minimal digging of the bed would be encouraged. In some fossil areas which are often visited, good specimens may be found on the surface.

Students should come equipped for the usual outdoor field excursion as well as bringing a rock chisel, hammer, newspaper or zip lock bags (to wrap specimen) and a carry bag for the specimen. See the video on finding specimen in shales:

 Online Video: How to collect plant fossils.
Go to https://youtu.be/0Hlmx3iBhf0

On returning to the classroom, a detailed excursion report based on the objectives of the excursion should be submitted to reinforce the concept that field research is always followed-up by laboratory and literature research. An activity similar to 8.1 could also be carried out.

- Make replicas of a fossil or other plant or hard-shelled animal specimen - this could be an extended group workshop to produce plaster replicas for the school collection and the students' personal use. It will be a very messy activity and so aprons and sufficient desk-covering will be needed.

- Carbon print simulation. Many plant fossils are often found as carbon prints within a shale or even fine sandstone. These can be simulated by having students:

    1. Make a slab of plaster by pouring Plaster of Paris into a framework of cardboard (e.g. a circular strip about 1 cm high stuck onto a cardboard base with tape).

    2. Before setting, place a leaf obtained from the schoolyard or home (a herringbone fern is best) onto the plaster's surface and allow it to be set within the plaster.

    3. The next day, hold the plaster slab upside down over a gentle BLUE flame of a Bunsen Burner and gently singe or carbonise the plant.

    4. When cool, carefully brush away any excess carbon on the surface of the plaster (but NOT from the leaf).

12.5 Answers to Multichoice Questions

Q1.B Q2.B Q3.B Q4.B Q5.B Q6.B Q7.D Q8.A Q9.D

12.6 Some Suggestions for the Review Questions

1. **What was the theory of spontaneous creation? How was it experimentally disproved?**
    The theory stated that life could arise from non-living matter e.g. maggots from rotten material. It was disproven by Pasteur's experiments.

2. **Compare and contrast the organelles of plant and animal cells.**
    Both have the same basic organelles such as a nucleus, nucleolus, cell membrane, vacuoles, cytoplasm and mitochondria but plant cells also have chloroplasts and cell walls and often larger vacuoles.

3. **Since the Cambrian Period, mosses, ferns, conifers and angiosperms have evolved.**

    (i) **Identify ONE advantage the terrestrial environment offered the first land plants.**
    Easier access to atmospheric gases and a variety of reproductive types.

    (ii) **Using examples, explain how plants had to change to overcome environmental difficulties as they moved from aquatic to terrestrial habitats.**
    Plants needed to adapt to loss of water through tissue e.g. simple aquatic algae evolved to land plants with a water-proof cuticle.

    See: https://www.ck12.org/biology/plant-evolution/lesson/Plants-Adaptations-for-Life-on-Land-MS-LS/

4. **Detail the main ways by which ancient organism can be preserved.**
    Organisms can be preserved in ice/permafrost, amber, tar, salt and as impressions and hard body mineral replacement or alteration.

5. **Why are fossils rare to find? Give some reasons why fossils may not be found in a locality.**
    Many reasons such as no life existed there originally; remains were exposed and decayed, removed or eaten; covered by sediment or lava; disturbed by land movement; weathered and eroded with the surrounding rock.

6. **Why is taxonomic classification necessary? Outline the binomial system of Linnaeus.**
    There are so many organisms that an unsystematic study would be difficult and take too long. Classification allows groups of similar organisms to be studied as a group. Carl Linnaeus, in his *Systema Naturae* (1735) and subsequent works, suggested three kingdoms, divided into *classes*, and they, in turn, into *orders*, *genera* (singular: *genus*), and *species* (singular:

species), with an additional rank lower than species. Using his binomial system, every organism could be classified with two names – genus and species.

7. Suggest a possible environment for each of the following fossils:

    a. *Dicroidium* – cold, wet, rainforest of ferns and other simple plants.
    b. *Lovenia* – a sea urchin living in shallow sea margins.
    c. *Favosites* – a coral living in warm, shallow tropical seas.
    d. *Pachypleurosaurs* – small marine swimming dinosaur.
    e. *Eohippus* – small, dog-sized horse with toes living in inland marshes.

8. Outline briefly the steps which would be taken

    a. before and
    b. after

    either undertaking a water sampling trip to collect organisms from a stream environment

    or

    mounting a personal hunt for fossils in a selected area.
    Depends upon the nature of the expedition but one would have to look at items needed for sampling/collecting, environmental testing, photography or sketches, transportation, communications and all necessary food, water and shelter.

9. Studies of a particular group of trilobites show a number of significant changes in general body structure over time from oldest to latest due to evolution. These changes include a(n):

    - decrease in the number of thoracic segments (not lobes)
    - enlargement of the pygidium
    - increase in ornamentation of and spines at the end of the pleura
    - shortening of the glabella with fewer segments
    - an increase in eye size

    Given the nature of the ancient form (shown below)

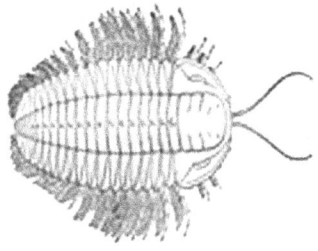

<u>Draw</u> a simple sketch of a possible species which may have later evolved from this species (Note: this is a good exercise with no exact correct answer)

Students have fun with this question, especially if it is drawn on large A3-sized paper and then displayed to all with a vote for the best drawing and most accurate.

10. A shale in a certain location was found to contain the following plant fossils:

Dicroidium dubium, D.zuberii, D. elongate and *Taeniopteris sp.*

Upon returning to base and consulting reference books, the following table was found:

GENUS & SPECIES	GEOLOGICAL PERIODS				
	Late Permian	Triassic			Early Jurassic
		Early	Middle	Late	
Glossopteris species	───────	───		───	
Voltziopsis sp.		─────			
Danaeopsis hughesii		───		─────	
Dicroidium brownii		──			
D. zuberi		─────			
D. dubium		──			
D. narabeenense			─────		
D. lancifolium			───────		
D. elongata			───────────		
Pleuromeia sp.		──			
Pachydermophyllump			───────		
Taeniopteris sp.			─────		

Use this table to determine the age of the shale which contained the fossil plants. Give the scientific laws or principles which enabled your conclusion.

Early Middle Triassic as this is the only time when these fossils could have lived together and been preserved in the rock.

12.7 Reading List

Alvarez LW, Alvarez W, Asaro F, Michel H.V. (1980). Extraterrestrial cause for the Cretaceous-Tertiary extinction. *Science* 208 (4448): pp.1095-1108. Bibcode: 1980Sci.208.1095A.doi:10.1126/science.208.4448.1095.PMID 17783054.

Azzaroli, A. (1998). The genus Equus in North America. *Palaeontographla Italica* 85: pp. 1-60.

Beerling, D. *The Emerald Planet: How Plants Changed Earth's History*. Oxford, UK: Oxford University Press, 2007.

Bell, E., Boehmke, P., Harrison, M. and Mao, W.L. (2015). *Potentially biogenic carbon preserved in a 4.1 billion-year-old zircon.* Proceedings of the national academy of science. November 24, 2015 112 (47) 14518-14521.

Cairns-Smith, Alexander Graham (1990), Seven Clues to the Origin of Life. Cambridge UP, ISBN 9780521398282

Cavalier-Smith, T. (2007). A revised six-kingdom system of life. *Biological Reviews* 73 (3): pp203-266.

Chapin, F. S., Matson, P. A. et al. (2002) Principles of Terrestrial Ecosystem Ecology. New York, NY: Springer Science+Business Media Inc.

Chapin, F. S., Matson, P. A. et al. (2002) Principles of Terrestrial Ecosystem Ecology. New York, NY: Springer Science+Business Media Inc.

Choi, C.Q. (2013). *Asteroid Impact That Killed the Dinosaurs: New Evidence.* Live Science Contributor. February.
http://www.livescience.com/26933-chicxulub-cosmic-impact-dinosaurs.html

J. Cook, et al. (2013) Quantifying the consensus on anthropogenic global warming in the scientific literature. *Environmental Research Letters* Vol. 8 No. 2, (June); DOI:10.1088/1748-9326/8/2/024024.

Cowen, Richard. (1994). *Tracking the Course of Evolution* - The K-T Extinction. In History of Life, 2nd. Edition. Boston, Massachusetts: Blackwell Science 460 pp.

Dodd, Matthew S.; Papineau, Dominic; Grenne, Tor; Slack, John F.; Rittner, Martin; Pirajno, Franco; O'Neil, Jonathan; Little, Crispin T. S. (1 March 2017). *Evidence for early life in Earth's oldest hydrothermal vent precipitates".* Nature. 543 (7643): 60-64. doi:10.1038/nature21377.

Endangered Species International (2011).*The Five Worst Mass Extinctions* http://www.endangeredspeciesinternational.org/overview.html

EPA (the US Environmental Protection Agency) *Causes of Climate Change.* Last updated on 21/7/2015.
http://www.epa.gov/climatechange/science/causes.html

Fortey, R. (1999). *Life: A Natural History of the First Four Billion Years of Life on Earth.* Vintage. pp. 238-260. ISBN 978-0-375-70261-7.

Gon III, S. (2015). A Guide to the Orders of Trilobites - A website devoted to understanding trilobites. http://www.trilobites.info/

Gribbin, J.R. (1982). *Future Weather: Carbon Dioxide, Climate and the Greenhouse Effect.* New York:Penguin.

Imbrie, J. & Imbrie, K.P. (1979). *Ice ages: solving the mystery.* Short Hills NJ: Enslow Publishers. ISBN 978-0-89490-015-0.

Ivanov, M., Hrdlickova, S & Gregorova, R. (2001). *The Complete Encyclopedia of Fossils*. Praha, Solvakia: Rebo. pp.312.

Kolbert, E. (2014). *The Sixth Extinction: An Unnatural History*. New York: Henry Holt and Company. ISBN 978-0805092998.

Lambert, D. and the Diagram Group. (1985). *The Field Guide to Prehistoric Life*. New York: Facts on File Publications.

McLoughlin, S. (2013). The fossil flora of Dinmore. *Australian Dinosaurs* Issue 10 pp 40-49 Febuary.

Moore, R.C., Lalicker, C.G and Fischer, A.G. (2004). *Invertebrate Fossils*. New York: McGraw-Hill. ISBN 10: 8123911394

NASA (2011). *NASA Research Shows DNA Building Blocks Can Be Made in Space* https://www.nasa.gov/home/hqnews/2011/aug/HQ_11-263_Meteorites_DNA.html

NewWorld Encyclopedia – Trilobites. (2008). http://www.newworldencyclopedia.org/entry/Trilobite.

Prothero, D. R. and Shubin, N.(1989). The evolution of Oligocene horses. *The Evolution of Perissodactyls*. pp. 142-175. New York: Clarendon Press.

Rey L.V, Holtz Jr T.R. (2007). *Dinosaurs: the most complete, up-to-date encyclopedia for dinosaur lovers of all ages*. New York: Random House.

Royal Society (UK) & the US National Academy of Sciences (US). (2014). Climate Change – evidence and Causes: An overview from the Royal Society and the US National Academy of Sciences, February. ISBN 978-0-08-020409-3. http://dels.nas.edu/resources/static-assets/exec-office-other/climate-change-full.pdf

Scoville, Heather (2018). *Early Life Theories - Panspermia Theory*. thoughtco.com/early-life-theory-of-panspermia-theory-1224530.

Scott, P.T. (1984). Some Common Invertebrate & Plant Fossils. . *Science & Ag. Bulletin*. 6 (3) Tamworth NSW, NW Region, NSW Dept. Ed., pp 46-66.

Scott, P.T. (2016). Fossils – Life in the Rocks. Brisbane; Felix Publishing. ISBN: 978-09946432-4-7

Scott, P.T. (2018). Beyond Planet Earth. 2nd Edition. Brisbane: Felix Publishing. ISBN: 978-0-9946432-4-7

Szostak, Jack W. (2009). *Origins of life: Systems chemistry on early Earth. Nature.* 459 (7244): 171-72. doi:10.1038/459171a. PMID 19444196.

Tanner LH, Lucas SG & Chapman MG. (2004). Assessing the record and causes of Late Triassic extinctions. *Earth-Science Reviews* **65** (1-2): pp. 103-139. doi:10.1016/S0012-8252(03)00082-5.

Wallis, N.K.; Wickramasinghe N.C. (2004). *Interstellar transfer of planetary microbiota*. Mon. Not. R. Astron. Soc. 348: 52-57.

White, M.E. (1988). *Australia's Fossil Plants*. NSW: Reed Books. ISBN 0 7301 0259 9

Wooldridge, S. A. (2008). *Mass extinctions past and present: a unifying hypothesis*. *Biogeosciences* **5** (3) June: pp. 2401-2423. doi:10.5194/bgd-5-2401-2008.

Chapter 13: Energy and the Earth

13.1 Theme

Mechanical energy has been defined as the ability to do work i.e. move masses through distances using a force. This mechanical energy could easily be converted to heat energy by friction and soon it became apparent that energies were being changed from one form to another with matter being involved as another source of energy.

13.2 Rationale

How energy can be changed, transferred and obtained from matter is essential to understanding many of the processes of all of the spheres of the Earth. Energy interaction between the Sun and all life upon the Earth is a major component of the Earth as a system and of the development and maintenance of biota on the planet.

13.3 Notes on the Practical Work (see PRACTICAL MANUAL 1.)

<u>Experiment 13.1</u>: Law of Conservation of Energy

This is a neat experiment to show, within some acceptable error, that energy is conserved when being changed from mechanical energy to heat.

<u>CONCLUSIONS:</u>

1. What was the change in temperature of the lead?
2. How did this agree with the theoretical value (of approx. 7.6 Celsius degrees)?
3. What was the error of this experiment expressed as a percentage error (i.e. absolute error/theoretical value x 100/1 %)?
4. Considering the errors, did the Law of Conservation of Energy operate? Discuss.
The answers for these four questions will depend upon the experiment. It usually works so there should only be a small error and the law should be demonstrated.

5. Account for this by listing some of the errors which may have occurred.
The most obvious error will be the transfer of heat between the inside of the tubing and the environment. PVC is a relatively poor conductor of heat so the error should be small. There may also be an error in reading the temperatures so it is vital that the thermometer is quickly pushed well into the lead slot (CARE that the glass is not broken) before too much heat escapes to the air and beaker.

6. Are there ways of improving this experiment? Suggestions?
The tube could be rotated for many more times (in a relay perhaps?) and a metal thermometer such as used in cooking could be inserted into one of the stoppers of the tube to measure the shot directly instead of having to empty it out.

Additional Research:

The students could research the life and contribution of Sir Benjamin Thompson, Count Rumford who noticed the conversion of mechanical energy to heat during his activities in boring out cannon. See:

https://www.britannica.com/biography/Sir-Benjamin-Thompson-Graf-von-Rumford

http://www.physik.uni-leipzig.de/~kroy/materials/rumford.htm
(Quote from Count Rumford about his work)

Experiment 13.2: Anaerobic and Aerobic Respiration

These two experiments show the production of carbon dioxide from anaerobic respiration by the fermentation of yeast and the aerobic respiration of Humans. In PART A, it is important to use a fresh yeast supply and to maintain the water temperature at the optimum. Initially the water should be about 30^{0C} to 40^{0C}. The limewater also should be clear to start with. In PART B, it is important to remind the students not to breathe in when placing the drinking straw into the limewater and to blow gently to prevent back splashing.

See: http://www.rsc.org/learn-chemistry/resource/res00000470/fermentation-of-glucose-using-yeast?cmpid=CMP00005115

If limewater is accidently ingested, there should be no major problem at that concentration – give the student plenty of water to drink. See:

https://www.nwmissouri.edu/naturalsciences/sds/l/Limewater.pdf

PART C is a good introduction to bread making.

CONCLUSIONS – ALL PARTS:

1. Was there any change in the temperature of the yeast mix in the conical flask in PART A? If not explain why. Compare this to any temperature changes PART C – was there any change to the temperature in the ball with the yeast? Explain.
There should be an increase in the temperature as respiration is exothermic but it might not be too obvious in part A due to loss of heat to the flask and surroundings. Internal heat in the dough should be noticed as the dough is also a good insulator.

2. Why was a flour ball made without yeast?
This was used as a control experiment to see if it was the effect of the yeast and not just the effects of the dough.

3. What did the change in limewater in both PART A and PART B show?
The limewater should become cloudy due to its reaction with carbon dioxide from the respiration to form insoluble calcium carbonate.

4. Why were separate test-tubes of limewater placed nearby the experiments? Compare their colour with that of PART A and B.
This is another control to compare the reaction of atmospheric carbon dioxide.

5. Was there any change in the colour of the limewater which was set aside? If so, why did that occur? What was the purpose of this test-tube?
There should be a slight reaction because of the carbon dioxide within the air.

6. Describe the smell of fermentation.
There should be a pleasant, sweet smell becoming more pungent with alcohol formation.

7. List all of the errors which could occur in this experiment.
The main error will be in the freshness of the yeast and the temperature at which the culture is maintained. If kept overnight in a cool place the yeast reaction will stop. Also, any heat generated in PART A may be lost to the surroundings.

8. Write a general statement about the two types of respiration observed.
This should show that the student understands the idea with the concept of free oxygen being used in aerobic conditions and glycolysis with anaerobic respiration. See:

https://www.bbc.com/bitesize/guides/zm6rd2p/revision/2

Experiment 11.3: Energy and the Water Cycle

This is a clear and simple model of the Water Cycle and the use of energy between the phases of water. Whilst a simple activity, students should be alerted to the fact that there are many observations to make such as the water condensation on the bottom of the watch glass is clear and not coloured representing freshwater coming up from the 'salt' water below. One could use salt water but the change to fresh water is difficult to detect. Also notice the rapid decrease in the shape of the ice as it cools the warm vapour coming off the warm water (and the warm beaker).

CONCLUSIONS:

1. What happens on the underside of the watch glass inside of the beaker? What colour is the substance which forms there? Explain.
The water vapour condenses as clear-coloured liquid water as the dye is left behind.

2. What does this substance represent?
The clear water condensation represents condensation in the atmosphere as clouds.

3. What happens to the ice cubes? Why? What do they represent in this model?
They melt as they absorb the heat of the water vapour changing to liquid. They represent the cooler air of the atmosphere at height.

4. Comment on the use of this model to show some of the features of the Water Cycle.
It is a simple but very good model of part of the Water Cycle.

13.4 Other Activities

1. Show how mechanical energy can be changed to heat by rubbing hands together, striking a match, making fire using firesticks (if competent) see:

   https://www.wikihow.com/Start-a-Fire-with-Sticks

   https://outdoors.stackexchange.com/questions/218/how-can-i-start-a-fire-by-rubbing-two-sticks-together

2. Show other changes of energy e.g.

   a. Heat to electricity – use a thermocouple such as copper and iron wire ends (bared) twisted and heated strongly over a Bunsen flame whilst the other ends are connected to a galvanometer.

   b. Electrical to light (lamp), heat (connect steel wool to a power source with fuse and watch the sparks), sound (electric buzzer etc.), mechanical (electric motor).

3. Simple distillation of saltwater using laboratory equipment or construct a simple solar 'still, see:

   https://worldwaterreserve.com/potable-water/purification/how-to-make-a-solar-still/

13.5 Answers to Multichoice Questions

Q1.C Q2.D Q3.C Q4.B Q5.D Q6.C Q7.B Q8.A Q9.A Q10.C

13.6 Some Suggestions for the Review Questions

1. **Explain the difference between energy and power. What units are used to measure both?**
   Power (in watts) is the amount of energy (in Joules) used or taken in per second.

2. **Most coal-fired power stations only have an efficiency of about 45%. If the type of coal used produces 2500 kJ/kg when burned, how energy would be produced if one metric ton (=1000kg) of coal is used?**

If 100% efficient, then the station would produce 2500 x 1000 kJ, but at 45% efficient this would only be 1125000 kJ (1125 MJ)

3. **During the day, a 50,000-litre swimming pool is heated by the Sun so that the water temperature is raised from 20°C to 25°C. Assuming that 1 litre weighs 1 kg. and that the specific heat of water is 4.186 joule/gram °C, how much heat energy is involved?**
The water in the pool weighs (50,000 x 1000) and $Q = c\, m\, \Delta T$ or
50000000 x 4.186 x (25-20) = 1046500000 joules (1046.5 MJ)

4. **Explain the difference between exothermic and endothermic reactions giving two examples of each.**
An exothermic reaction is one which gives out energy e.g. respiration whereas an endothermic reaction takes in energy e.g. photosynthesis.

5. **Define each of the following terms:**

   a. **latent heat** – hidden heat absorbed or given out when there is a phase change
   b. **specific heat** - the amount of heat per unit mass required to raise the temperature by one degree Celsius.
   c. **Transpiration** – the removal of water through the stomates of leaves in plants.
   d. **radioactive decay** – the breakdown of heavier isotopes of elements to simpler isotopes with the liberation of nuclear radiation and particles.

6. **Compare and contrast:**

   a. **monosaccharides and disaccharides** – monosaccharides are single sugars whereas disaccharides are double sugars.

   b. **ATP and ADP** – ATP is Adenosine Triphosphate, made up of an adenosine molecule and three inorganic phosphates. When one of these phosphates is removed, the energy is produced with the resulting compound ADP, Adenosine Diphosphate.

   c. **aerobic and anaerobic respiration** – aerobic respiration is the breakdown of sugars using free oxygen whereas anaerobic respiration is the breakdown of sugars without oxygen.

   d. **conduction and aduction (Error: should be advection)** – conduction is the simple passage of heat along any medium that will carry heat whereas advection is vertical the transfer of heat or matter by the flow of a fluid.

   e. **permeable and impermeable** – permeable materials allow the passage of water through them but impermeable do not.

f. **artesian and sub-artesian** – artesian water is that which is confined in aquifers below the surface under pressure and will come up to the surface without pumping. It is usually hot, often rich in salts and sometimes natural gas. Sub-artesian water is not confined under much pressure and usually has to be pumped to the surface.

7. **What is hydrolysis? Why is it important to the cells of organisms?**
   Hydrolysis is the breakdown of large molecules using water as a reactant e.g. the breakdown of ATP to ADP in cells.

8. **What is meant by the Earth's Energy Budget? Why do scientists go to considerable effort to calculate the values expressed in the Earth Energy Budget?**
   This is the balance between the energy received from the Sun and the energy lost back into space. It is important to measure if the Earth's surface is cooling or warming.

9. **Hawaii has active volcanoes yet it is in the middle of the Pacific Plate. Most volcanoes occur around the edge of plates. Explain.**
   Hawaii lies over a 'hot spot' or mantle plume of heat coming up to near the surface. This melts the basaltic crust below the ocean and forms volcanic islands such as those at Hawaii.

10. **Use the internet to find out how much energy is used by the average home. Discuss ways by which energy uses in the home could be reduced?**
    This is an important exercise. Students could keep a diary of what appliances are used on a regular basis each day and then calculate the amount of energy used. Details of an appliance should be on the plate at the base or on the back or the student could look up the internet to find out these values. See:

    https://www.energuide.be/en/questions-answers/how-much-energy-do-my-household-appliances-use/71/

    Remember that power (in watts) equals energy (in joules) per second so the amount of time that these appliances are used must also be known.

13.7 Reading List

Kushnir, Y. (2000). Solar Radiation and the Earth's Energy Balance. Published on The Climate System, a complete online course material from the Department of Earth and Environmental Sciences at Columbia University.

Lindsey, Rebecca (2009). *Climate and Earth's Energy Budget*. NASA Earth Observatory.

Lofts, G; O'Keeffe D; et al. (2004). "11 — Mechanical Interactions". Jacaranda Physics 1 (2 ed.). Milton, Queensland, Australia: John Willey & Sons Australia Ltd. p. 286. ISBN 0-7016-3777-3.

Nelson, David L., Cox, Michael M. (2013): Lehninger Principles of Biochemistry. Sixth ed. New York: W.H. Freeman and Company, 2013.

National Oceanographic and Atmospheric Administration (NOOA) (2006) *"Hydrologic Cycle - Northwest River Forecast Center*.

Peixoto, J., and Oort, A. (1992). *The ocean-atmosphere heat engine*. In Physics of Climate (pp. 365-400). Woodbury, NY: American Institute of Physics Press.

Smil, Vaclav (2008). *Energy in nature and society: general energetics of complex systems*. Cambridge, USA: MIT Press. ISBN 0-262-19565-8.

The United Nations (2007). *Melting Ice—A Hot Topic? New UNEP Report Shows Just How Hot It's Getting*. United Nations Environment Programme (UNEP). 2007-06-04.

Tipler P. (2004) Physics for Scientists and Engineers: Mechanics, Oscillations and Waves, Thermodynamics (5th edition). New York: W.H. Freeman.

Walding, Richard; Rapkins, Greg; Rossiter, Glenn (1999). New Century Senior Physics. Melbourne, Australia: Oxford University Press. ISBN 0-19-551084-4.

Chapter 14:   Energy and the Sea and Sky

14.1 Theme

Energy transfer and differences are also major factors in driving the great process, especially currents in the atmosphere and hydrosphere. The nature of the composition of the atmosphere, its carbon dioxide component and its particulate nature also affect the amount of energy being retained or reflected from the planet assisting in the Greenhouse Effect. Energy transfer is also important in the biosphere within cells and entire populations of communities.

14.2 Rationale

Energy pervades all things on the planet; the sea and sky and all of the living things which live on the surface. Energy changes affect all of the great processes of the Earth and the relationships between the energy transfer in the abiotic and biotic worlds is important to understanding how life is maintained on the planet.

14.3 Notes on the Practical Work (see PRACTICAL MANUAL 1.)

Experiment 14.1: Radiation and the Inverse Square Law

This is a simple activity showing one of the most important laws of radiation of energy. It is important to keep the room darkened and to use a narrow source of light.

CONCLUSIONS:

1. What is the shape of the graph? What mathematical relationship does it show?
If external light is kept to a minimum, the graph should be an inverse straight line.

2. Does this graph show the Inverse Square Law for electromagnetic radiation?
The experiment usually shows the relationship to be true.

3. What would be the errors in this experiment? How could they be removed or kept constant?
Errors are due to the additional light coming into the room. This could be measured initially and subtracted from the results or (with care) the room could be made completely dark. Also there could be insensitivity of the light meter so one should be chosen to give a full-scale deflection for the final position.

4. Explain, using a suitable diagram why the Inverse Square Law works.
  See:
     http://hyperphysics.phy-astr.gsu.edu/hbase/Forces/isq.html

Experiment 14.2: Modelling the Greenhouse Effect

This experiment can be difficult to manage depending upon how it is set up and could be set up as a demonstration. Ideally, one needs a rectangular sealed bottle and a shallow transparent tray to cover the side of the bottle when it is lying flat. It has been tried using a flat, plastic dish and an empty, rectangular cordial or juice bottle which has had a small hole made in its plastic top. Through this a thermometer was pushed and sealed with putty. The tray of water on top of the bottle represents the atmosphere and its ability to absorb heat radiation. The lamp is used to change the amount of light falling upon the 'atmosphere' of the bottle. PART C gave some problems when trialling the experiment as CO_2 was made using a simple generator using a fruit salt/water mixture connected to the bottle by a plastic tube. A better supply of dry carbon dioxide would be more appropriate such as from a commercial CO_2 gas cylinder or Kipps generator.

CONCLUSIONS:

1. Was there any change in temperature in the Control Experiment? What does indicate?
All errors accounted for there should be a rise in the temperature.

2. Was there any change in temperature in the bottle in Part A? What does this model suggest about the temperature effects due to the distance of Earth from the Sun?
Any increase in temperature would be due to the extra light at closer distances.

3. Was there any difference in temperature when the 'surface' was changed from white to black? Explain.
This replicated the lack of albedo from darken surfaces so the temperature should be higher with a darkened surface.

4. Did the presence of additional carbon dioxide gas in the bottle change the temperature compared to the bottle with ordinary air? Explain.
In theory and in the trials made, there was a slight increase due to the absorption of heat by the carbon dioxide.

5. This is a very basic model for what has been termed the Greenhouse Effect but there would be many problems with such a model. Suggest any problems with this model and if possible, suggest some improvements (some research may be needed).
There are a number of possible errors in this model due to its simplicity and the very task of modelling the greenhouse effect i.e. the absorption of light energy by carbon dioxide after the light has been reflected off the ground. Good luck!

6. Could there be any other factors other than the three tested here that could change temperature through a Greenhouse Effect? Suggest how this model may be modified to test these hypotheses.
Other factors may include the particulate nature of the atmosphere – injecting smoke may show some drop in the temperature or the use of filters to replicate the effects of clouds.

Experiment 14.3: Convection Currents

This is a simpler experiment which is easy to set up and operate. It is important to have a large beaker, at least 800 ml, and a single, large crystal of potassium permanganate placed on the side at the bottom of the beaker. The burner must be placed just under the crystal and set on a medium blue flame. Observations should start as soon as the burner is placed below the beaker. Most students will miss the condensation of water vapour on the outside of the cold beaker as soon as the flame meets the glass. A good challenge!

CONCLUSIONS:

1) **What happens to the water:**

   (a) immediately above the crystal - it will rise, taking the colour with it.
   (b) at the surface of the water - it spreads across the surface.
   (c) on the opposite side of the beaker from the burner (cool side)? - it will start to sink.

2) **What is a convection current?** - the movement of a hot fluid rising up when hot and falling when cold, often giving circular currents.

3) **How do convection currents operate in the world's oceans?**
Water is warmed by the Sun in the tropics and spreads out to colder climates where it sinks and forms a counter current to rise again due to the surface of the sea floor to be warmed again.

4) **Given an example of one of the great convection currents in the ocean by naming:**

   (a) a warm ocean current - the Gulf Stream
   (b) a cold ocean current - the Peru/Chile (Humboldt) Current

Experiment 14.4: Photosynthesis

An excellent experiment but may take some time and the apparatus needs to be in sunlight for a good time. Elodia is excellent and can be obtained from Aquarium shops. Use ordinary tap water which will contain enough dissolved air to supply the carbon dioxide required by the plant. Care must be taken when inverting the test-tube full of water over the beaker as atmospheric air may leak in. This can be prevented by filling and inverting the tube under water in a large container and then sliding a square of wet cardboard over the top of the tube. This should stay in place by air pressure until it is placed over the spout of the funnel in the apparatus. All variations of the experiment must be done in direct sunlight for maximum effect

CONCLUSIONS:

1) Was there any difference between the number of bubbles over the total time for each of the different variations? If so:

    (a) which one gave the greater number of bubbles (gas volume)? Explain.
    (b) which one gave the least amount of bubbles? Explain.

If the experiment works well then, the experiment initially set up should give the best results and the one with a green filter should be the least. If a control is set up which is in a darkened cupboard then this should show the least result. The red cellophane allows red light to pass through and it is the reddish end of the spectrum that plants use. Green cellophane acts only as a reducer of the light as plants are green and reflect that colour.

2) If the gas was tested, what gas was it? Why?
If there is sufficient gas volume, a glowing splint should reignite showing the presence of oxygen. This may be difficult depending upon the amount of gas. At least a third of a test-tube is needed so several versions could be started early and kept in the sunlight over several days.

3) What other factors could be changed to determine how photosynthesis works?
Try using blue light as it is also absorbed into chlorophyll. Also try varying the temperature of the water from room temperature to cold water. Use demineralised water which has also been boiled and cooled to remove any air. Pour a layer of cooking oil on top of the water in the beaker to prevent any CO_2 from being absorbed from the air.

4) What errors could have occurred in the experiment?
These include insufficient time, light and amount of plant to get good volumes of oxygen, student error in counting bubbles and air getting into the tube when inverting the test-tube into the beaker.

5) Write a general statement about photosynthesis in plants and its requirements.
A general statement that plants need green chlorophyll, carbon dioxide, water and sufficient light energy of the right colour (red & blue) and an optimum temperature to allow photosynthesis to work.

Experiment 14.5: Population Growth of a Bacteria Colony

This activity was chosen for simplicity and convenience. Teacher with good biological skills and equipment, especially a good biological microscope with video camera might like to reproduce the bacterial cultures as a demonstration.

If students have individual tablets or laptop computers, they are able to do the activity individually, otherwise a data projector or smartboard may be used with the website and as a demonstration with group counting of the number of bacteria at each pause. It will only be practical to measure the numbers over the first few time pauses as the screen soon becomes full of hundreds of bacteria. In addition, the video can only show the growth phase. Eventually the numbers will flatten out as the optimum population is reached and then decline as they begin to die due to lack of nutrient and waste accumulation.

CONCLUSIONS:

1. Describe the shape of the final graph.
It will be a sharp J-curve due to exponential growth.

2. Mathematically, what relationship does this graph show?
An exponential relationship i.e. $y = a^x$

3. Is it consistent with the usual logistics curve for populations in nature?
It is consistent with the growth-rate part of population growth.

4. How could this (original) experiment be expanded so that the bacteria can reach their climax population?
This would be difficult but one would need a larger view or do the experiment in real life and measure the size of any bacterial growth on an agar plate.

5. In the Petrie dish, what factors would limit the growth of the bacteria?
The amount of nutrient and space. Also, any wastes produced will also limit growth as they reach their optimum population number.

14.4 Other Activities

1. Repeat the Inverse-square Law experiment for infra-red using a heat lamp.

2. Extract chlorophyll from the leaves of a non-waxy plant by boiling it in a beaker of alcohol set in a water bath and cautiously heated.

3. Soak the leaves from this last activity in tincture of iodine to see the starch (blue) produced by photosynthesis.

4. Use the extracted tincture of chlorophyll in a paper chromatography activity by soaking the ends of tall strips of blotting paper into beakers with a 1 cm layer of tincture.

5. Cover leaves on a growing plant with dark cardboard or paper (or aluminium foil) and then repeat activity 2 and 3 (above).

6. Obtain small, cylindrical pill containers with plastic lids. Make a hole in the lid close to the edge and push an eyedropper through it. Fill the eyedropper with very warm water dyed red. Fill the pill container with very cold water and carefully replace the lid with the eyedropper without squeezing the bulb. Turn the container on its side with the eyedropper nozzle on the bottom side. Gently squeeze the eyedropper bulb and watch the warm, red water rise to the top of the tube and flow along its top. Repeat this activity

using very cold water dyed blue in the eyedropper and warm water in the container. This time, turn the container on its side with the eyedropper at the top. Squeeze the bulb and watch the cold water sink.

14.5 Answers to Multichoice Questions

Q1.C   Q2.C   Q3.B   Q4.B   Q5.D   Q6.D   Q7.A   Q8.B   Q9.C   Q10.D

14.6 Some Suggestions for the Review Questions

1. **Explain several of the ways that life on the Earth's surface is protected from harmful emissions from the Sun. Give detail of each of the potential hazards and how they are stopped or reduced.**
   e.g. magnetosphere prevents charged particles from the solar wind, ozone layer filters out much of the UV radiation.

2. **What is albedo? What factors determine the amount of the Earth's albedo effect.**
   This is the ability of a surface to reflect light and other radiation. It depends upon the colour of the surface and its smoothness. Smooth, white surfaces such as snow and ice have the highest albedo.

3. **List the main natural greenhouse gases. Which one is potentially the most dangerous? Why? (some internet research may be needed).**
   Carbon dioxide gas, water vapour and methane gas are the most important greenhouse gases. Methane has many times the absorbing power than carbon dioxide (84 times) but it is not as common in the air as CO_2 and has shorter 'lifetime'. However, it is constantly being emitted by decomposing organic matter and manure and there are vast quantities stored in the frozen permafrost.

4. **Why are there changes in air pressure in different parts of the Earth? How is it measured and what are the units used to express changes in air pressure?**
   Differential heating causes air to rise over hot regions (giving low air pressure) and to sink over colder regions (high pressure). Air pressure is measured in hectopascals or millibars.

5. **Define each of the following terms:**
   a. **Hadley Cell** - a large-scale atmospheric convection cell in which air rises at the equator and sinks at medium latitudes, typically about 30° north or south.
   b. **Thermohaline system** – system of warm and cold ocean currents with differences in salt content and heat causing motion.

- c. **Blackman reaction** – is the stage wherein the chemical process of photosynthesis takes place without the use of sunlight. The reaction occurs in the stroma of the chloroplast.
- d. **Hess' Law** - states that regardless of the multiple stages or steps of a reaction, the total enthalpy change for the reaction is the sum of all changes.

6. **Compare and contrast winds and ocean currents by explaining how each is produced and any factors involved**

   Wind currents are due to differential heating of the land and sea surfaces and involve the lower part of the atmosphere whereas ocean currents are driven by the wind and heat from above as well as saltwater densities.

7. **In the Northern Hemisphere, hurricanes are intense low-pressure zones which rotate in a counter clockwise (anticlockwise) direction but in the Southern Hemisphere, a similar low called a Tropical Cyclone rotates in a clockwise direction. Explain.**

   This due to the balance between the force imparted by horizontal pressure differences, friction against the Earth's surface, and the Coriolis effect, which causes moving air to deflect to the left of their original direction in the Southern Hemisphere and vice versa.

8. **What would be the implications in the western Pacific if there was a prolonged positive value for the Southern Oscillation Index**

   Prolonged positive values of the SOI (above +7) are typical of a La Niña episode associated with stronger Pacific trade winds, warmer sea temperatures to the north of Australia and prolonged rain.

9. **Distinguish between each of the following**

    - a. **Autotrophs and heterotrophs** – autotrophs such as plants manufacture their own food but heterotrophs such as animals must obtain food from other organisms.

    - b. **Consumer and producer in ecosystems** – another version: producers, such as plants, make their own food and are then eaten by primary consumers that cannot produce their own food.

    - c. **Decomposers and carnivores**- decomposers primarily feeds on dead organisms or the waste from living organisms, breaking them down whereas carnivores feed on animals.

    - d. **El Niño and La Niña** - El Niño result from the weakened trade winds pushing warm water back toward the east towards the Americas with rain and drought in the western Pacific. La Niña is associated with stronger Pacific trade winds, warmer sea temperatures to the north of Australia and prolonged rain but less rain in the Americas.

    e. **Predator and parasite** – predators are animals which hunt other animals whereas parasites are organisms which live on and feed off other organisms to the detriment of the host.

    f. **Mutualism and cooperation between organisms**- Mutualism is the way two organisms of different species exist in a relationship in which each individual benefit from the activity of the other whereas cooperation is a similar relationship within one species.

    g. **Food web and food chain** - a food chain is a linear sequence of organisms through which nutrients and energy pass as one organism eats another whereas a food web consist of many interconnected food chains and are a more realistic representation of consumption relationships in ecosystems.

10. **Why are environmentalist concerned about the recycling of carbon, nitrogen and phosphorus?**
Because these elements make up much of the protein and other biological chemicals necessary for life.

14.7 Reading List

Ahrens, C. Donald, (2007). Meteorology today: an introduction to weather, climate, and the environment. Cengage Learning. pp. 296. ISBN 978-0-495-01162-0.

Australian Bureau of Meteorology (BOM): http://www.bom.gov.au/australia/

Brooks, H. E. (2004). *Estimating the Distribution of Severe Thunderstorms and Their Environments Around the World*. International Conference on Storms. Brisbane, Queensland, Australia.

Coakley, J. A. (2003). J. R. Holton and J. A. Curry, eds. *Reflectance and albedo, surface*. Encyclopedia of the Atmosphere. Academic Press.

Gnanadesikan, A., Slater, A., R. D. Swathi, P. S., and Vallis, G. K. (2005). The Energetics of Ocean Heat Transport. *Journal of Climate* 18 (14): 2604–16. Bibcode: 2005JCli...18.2604G. doi:10.1175/JCLI3436.1.

Heinemann, B. and the Open University. (1998). *Ocean Circulation*. Oxford University Press.

Lindsey, R. (J2009). *Earth's Energy Budget* and *The Atmosphere's Energy Budget*. In Climate and Earth's Energy Budget. Earth Observatory, part of the EOS Project Science Office, located at NASA Goddard Space Flight Center.

Earth's energy budget NASA
https://science-edu.larc.nasa.gov/energy_budget/pdf/ERB-poster-combined-update-3.2014.pdf

NASA GISS: Science Briefs: Greenhouse Gases: Refining the Role of Carbon Dioxide. www.giss.nasa.gov.

National Aeronautics and Space Administration (NASA): https://www.nasa.gov/

National Oceanic and Atmospheric Administration (NOAA): http://www.noaa.gov/

Nature Magazine (2013). *Deep ocean is a heat sink. Nature* volume 503, page 9 (07 November 2013).ISSN 1476-4687 (online). https://www.nature.com/articles/503009b

Perrino, Cinzia (2010). *Atmospheric particulate matter". Biophysics and Bioengineering Letters. 3 (1). ISSN 2037-0199.*

Pinet, P. R. (1996). *Invitation to Oceanography.* Eagan, MI: West Publishing Company. ISBN 978-0-314-06339-7.

Scott, P.T. (2017). Riches from the Earth. Brisbane: Felix Publishing. ISBN: 978-0-9946432-6-1

Scott, P. T. (2018). Through Sea and Sky. Brisbane: Felix publications. Print ISBN: 978-0-9946432-2-3

Sverdrup, K. A., Duxbury, A. C. & Duxbury, A. B. (2006). *Fundamentals of Oceanography,* McGraw-Hill, ISBN 0-07-282678-9

Thurman, Harold and Alan Trujillo. *Introductory Oceanography.*2004.p151-152. Prentice Hall; 5th edition (1996). ASIN: B000OIPPTO

Trenberth, K., Fasullo, J., Kiehl, J. (2009). Earth's global energy budget (draft copy). *Bulletin of the American Meteorological Society.* doi: 10.1175/2008BAMS2634.1

Chapter 15: Use of Resources and Energy

15.1 Theme

Humankind has always made use of the natural resources of the environment, from the earliest stone tools of the Stone Age, through the Bronze and Iron Ages to our modern era of synthetic materials. Development has been from simple application of natural materials to a complex development and use of manufactured materials and energies. This Chapter sets the scene for further chapters on renewable and non-renewable resources and attempts to explain that using resources is a natural part of Human activity. In later Chapters, it will be seen that the current emphasis is on using traditional non-renewable resources such as fossil fuels and nuclear materials for energy and many ores, stones and other natural resources for daily use. With such resources being depleted and with a rapid increase in demand with a growing population, there will have to be more emphasis on the use of renewable resources.

15.2 Rationale

It is important in an energy and material depleted world to think of both resources and energies as either being renewable or non-renewable. Non-renewable energies and resources are now seen as having a very limited future, whereas there is distant hope in the use of renewables.

15.3 Notes on the Practical Work (see PRACTICAL MANUAL 2.)

Experiment 15.1: Making Charcoal

An historical interesting experiment invoking the days before fossil fuels when wood and charcoal derived from it were the only sources of industrial energy and a raw material to make iron and steel. Charcoal is much lighter than wood and burns more efficiently, moreover it could be used in smelting metal ores to metals. Charcoal burners were important rural people in the Middle Ages and their work was considered essential to society. Unfortunately, the use of timber for construction, fuel and for making charcoal saw the demise of most of the large forests in Western Europe.

QUESTIONS:

1. **What changes were observed to indicate that a chemical reaction has taken place?**
   There should be a blackening of the wood and some vapours given off.

2. **Was the charcoal lighter than the wood?**
   The timber was not weighed before heating as only small amounts were used. This could be done with a larger amount in a bigger, sealed container with a hole in top. The charcoal should be much lighter than the wood.

3. **What happened when the lighted splint was put into the mouth of the test-tube during the reaction? Why?**

   In theory it should burn or ignite due to the presence of 'wood gas' – a mixture of hydrogen, methane, carbon monoxide, water vapour and other gases. The concentration may not be high enough in this experiment to show good results.

4. **Did the charcoal produced burn when held in a burner flame?**

   It should burn but remember to use tongs.

CONCLUSIONS:

1. **Write a general conclusion about the use of timber to make charcoal.**

   It is the material to use and the charcoal is lighter and burns better than dry timber.

2. **Apart from making charcoal for later metal smelting and heating, what else was manufactured?**

   Probably not apparent but the glowing splint may show that the gases produced are useful as a fuel also. Note that toxic carbon monoxide is also produced so it is unwise to make large amounts of charcoal or burn carbon-based fuels in confined stoves indoors.

RESEARCH: (Optional)

1. **Use the Internet to find out how charcoal is made commercially.**

   As in the ancient times it is made from burning timber in a confined or closed container. Large ovens are used instead of heaped piles of wood.
   See:

   http://www.madehow.com/Volume-4/Charcoal-Briquette.html

   http://islandblacksmith.ca/how-charcoal-is-made/

2. **What is coke in reference to metal smelting? Compare and contrast it to charcoal.**

   Coke is the coal equivalent of wood and charcoal. Coal is heated strongly in a closed container and various gases (as coal gas) are given off as well as other vapours which contain ammonium and other useful materials. Coke was once used in homes for heating as was the coal gas. Now coke is mainly used in smelting, especially of iron and steel. It is a hard, semi-metallic and porous-looking form of almost pure carbon.

3. **Research the uses of (activated) charcoal other than as a source of heat in smelters.**

   Charcoal is still used today as in filters, in medicine and in traditional methods of smelting and cooking.

15.4 Other Activities

- Collect information about life in the Middle Ages in a similar setting (rural or town) with an emphasis on energy and resource needs, especially for building, clothing, food and water.

15.5 Answers to Multichoice Questions

Q1.C Q1. A Q2.C Q3.D Q4.B Q5.A Q6.A Q7.B Q8.D Q9.D Q10.D
In the Second Edition, the second Question 1 has been deleted.

15.6 Some Suggestions for the Review Questions

1. It has been said that Stone Age people were the first conservationists. Is this true? Explain.
Probably they were and most peoples living in pure indigenous cultures still do. It has been said that modern civilizations have lost touch with nature and should go back to being with it rather than making use of it. Indigenous and ancient peoples knew about limited natural resources and had practices to conserve resources, especially food supplies.

2. What is flint? Why did ancient peoples consider it a most important material?
Flint is a hard form of precipitated silica which was mined by ancient Stone Age peoples and used as a sharp edge in arrow heads, axes and knives.

3. What were the advantages that people who practiced agriculture and animal husbandry had over hunter-gathering people?
They could settle in one place, conserving materials by building fixed dwellings and making use of the local land by planting domesticated plants and grazing domesticated animals. The territory needed for this was also much smaller than hunter-gatherers who relied upon roaming large areas of land in the hope that they would find food, water and shelter. Settling also gave more time for the development of social activities, art and cultural pursuits.

4. What were the advantages and disadvantages of using copper as a source of metal objects? Why was bronze the preferred material and how was it made?
Copper was relatively easy to obtain (the name is derived from the old name for Cyprus where it was first obtained), smelt and work. Bronze, an alloy with tin, was a harder, more durable metal. Tin was obtained from many areas including Cornwall in Great Britain and so it was part of the metal trading system of Europe.

5. How did earlier generations use renewable energy to power some of their requirements? Give some examples.
Timber as fuel, wind, water and the use of animals and humans were the only power sources.

6. Why did cities such as Salzburg become very important in the early history of civilization?

Salzburg (meaning salt town) was an important Celtic settlement because it had an abundance of salt, necessary to preserve food for the long European winters. Nearby was the town of Hallein from which the mineral name for salt, halite comes. Salzburg became an important trading city because of salt and because it is in the centre of many European trade routes.

7. What is the meaning of each of the following words:

(a) **plumbum** - Latin word for lead and source of the chemical symbol Pb
(b) **brass** - an alloy of copper and tin.
(c) **Neolithic** - the New Stone Age, the final part of the Stone Age about 12,000 years ago.
(d) **Smelting** - reducing oxides or other ores of metals to the raw metal, often using carbon and other materials with large amounts of heat.

8. Why was aluminium not used as a metal until the mid-19th century? (Internet search may be needed).

Aluminium, which makes up a large percent of the rocks and soils of the Earth, is such an active metal that it was very hard to extract by normal smelting. The first successful attempt was completed in 1824 by Danish physicist and chemist Hans Christian Ørsted and the metal was then considered to be more valuable than gold.

9. Why would charcoal be considered a renewable resource? Why would it probably not be used again as a large-scale fuel?

It is renewable because plantation timber could be used as its main resource, however a considerable amount would be needed to come up to the quantities of coal currently used in smelting.

10. Research some of the locations of natural oil and gas seepages in the world. What was used as a source of lighting and lubrication before the large-scale use of crude oil?

Ancient peoples used natural oil/tar seepages as a resource and as a lighting fuel. Seepages occur when oil from below is able to come up through permeable rock to the surface. Often natural gas comes to the surface and is ignited by friction or lightning such as in Azerbaijan. Tar from Trinidad was used for the British shipping trade and there are still many seepage sites around the world in California, eastern Canada and in the western Asia.

See:

https://walrus.wr.usgs.gov/seeps/where.html

https://sites.google.com/a/ingv.it/getiope/home/gas-seepage

15.7 Reading List

Cameron, R. & Neal, L. (2002). *A Concise Economic History of the World: From Paleolithic Times to the Present* 4th Edition. New York: Oxford University Press Inc. ISBN: 0195127056.

Derry, T. K. & Williams, T. I., (1993) *A Short History of Technology: From the Earliest Times to A.D. 1900*. New York: Dover Publications. ISBN: 0486274721.

McNeil, I. (1990). *An Encyclopedia of the History of Technology*. London: Routledge. ISBN 978-0415147927.

Rider, C. (2007). *Encyclopedia of the Age of the Industrial Revolution, 1700-1920*. Santa Barbara: ABC-Clio Greenwood. ISBN: 978-0-313-33501-3.

Street A. & Alexander W. (1998). *Metals in the Service of Man* (11th ed.). Penguin Books, London, ISBN 978-0-14-025776-2.

Chapter 16: Economic Minerals

16.1 Theme

Ores are minerals of economic importance which are used in the manufacture of many materials. They usually are complex chemical compounds which must go through many processes before the useful substance, often a metal, can be extracted and purified.

16.2 Rationale

Economic minerals along with coal and oil are the main features of the mining industry. A good understanding of their properties and usefulness is essential to the exploration geologist, geochemist and those within the mineral processing industry as well as amateurs who like collecting crystals.

16.3 Notes on the Practical Work (see PRACTICAL MANUAL 2.)

Experiment 16.1: Examination of Some Common Economic Minerals

This has the identical method of an earlier experiment looking at rock-forming minerals. It involves the same skills of observation and the use of Mohs' Scale to test the basic properties of minerals such as colour, streak, habit, lustre, cleavage, hardness, specific gravity, diaphaneity and chemistry.

As before, it is ideal if there is a class set of (say) 10 labelled boxes each containing numbered or lettered specimen of economic minerals. These may include: native copper, galena, sphalerite, pyrite, bauxite, malachite/azurite, haematite, chalcopyrite, pyrolusite, and barites.

Economic minerals can be purchased on line. Some sites are:

https://www.ebay.com.au/b/Collectable-Mineral-Specimens/3220/bn_2211448
buying minerals

http://www.mineral.org.au/dealers/dealers.html   Australian mineral sites

https://www.mindat.org/ds_8_High-end_Collector_Mineral_Specimens.html
International sites for minerals

http://earthtoleigh.com/documents/worksheets/5.1%20Minerals.pdf
 (A simple Multichoice test worksheet on minerals. Open as pdf and Save)

QUESTIONS:

1. **Why would colour be an unreliable property for identification? Give an example where colour might be confusing?**

   Minerals sometimes have variable colour and also may undergo some weathering which changes their colour e.g. haematite could be red, orange, grey and black.

2. **List each ore and state the metal(s) which can be extracted from it.**

   This depends upon the set provided. As given in the suggested list, the ores and metals would be:

   > native copper - copper
   > galena - lead
   > sphalerite - zinc
   > pyrite - iron
   > bauxite - aluminium
   > malachite/azurite - copper
   > haematite - iron
   > chalcopyrite - copper
   > pyrolusite - manganese
   > barites - barium

3. **Explain why it may be difficult to describe and identify ore specimens.**

   Often, they come as weathered specimens or are only finely spread throughout a rock.

4. **Are there any specific physical properties other than those main properties listed above which may be useful in quick location and identification of the ore?**

   Some ores such as magnetite are magnetic and some, like uraninite are radioactive and so they can be detected by geophysical devices (e.g. magnetometers and Geiger counters).

CONCLUSIONS:

List the code numbers, name and at least two <u>distinctive</u> properties of each mineral which will help in the quick identification of that mineral.
As with rock-forming minerals this helps in remembering the distinctive features of the minerals e.g.
Pyrite – metallic gold
Galena – grey heavy etc.

But this will depend upon individual preferences

RESEARCH: (Optional)

1. Use the Internet to find out the main mining regions which produce these ores and mark them on a national map.
   See:

   https://d28rz98at9flks.cloudfront.net/115542/115542_Operating_Mines_Map_2017.pdf
   (Australian sites)

   http://www.oresomeresources.com/resource/operating-mines-in-australia-interactive-google-map/
   (Australian interactive sites)

   http://www.burgex.com/the-united-states-of-industrial-minerals-map/
   (United States)

   http://www.europe-geology.eu/mineral-resources/mineral-resources-map/critical-raw-materials-map/
   (Europe)

2. Again, list the metals obtained from these ores and research their uses. Mostly they are sources of metals from the ores which are then used in a number of ways.
   See:

   https://quizlet.com/6311651/economic-uses-of-minerals-flash-cards/

   http://scienceviews.com/geology/minerals.html

   https://geology.com/minerals/   Very good.

Experiment: 16.2: Geochemistry of Ores and Economic Minerals

This can be a complicated experiment and with some larger, difficult classes the teacher may wish to do part or all of it as demonstrations. With good classes it is still complicated but a worthwhile experiment, linking geology with chemistry and showing the usefulness of chemical analysis in mineral identification.

CARE! Assume that all of the chemicals and specimen to be used are toxic so warn the students about ingesting material and handling chemicals. Use of the burners with a naked flame can be dangerous so students should wear non-flammable aprons if possible.

PART A: Flame Tests.

This is usually a lot of fun for the students and relates to their knowledge of fireworks. Ordinary paper clips can be bent so that they can be held at one (looped) end and the other (straight) end can be dipped in water and then into a SMALL container (e.g. small cylindrical pill bottle) of the metal salt. Care must be taken to thoroughly wash the end of the wire in water to prevent contamination.

The edge of the blue flame of the burner is used as it is relatively colourless and very hot. The colour of the first flame is to be noted. Some of the colours should be:

COPPER	rich green
SODIUM	yellow
BARIUM	pale green
STRONTIUM	crimson
LITHIUM	lilac

Other Activities in support of this experiment:

1. **Other colours** can be obtained from calcium salts (orange), potassium chloride (purple), copper chloride (blue), magnesium sulfate (white) and powdered iron (filings) makes many sparks (fun).

2. **Try mixing a couple of salts together** and get the students to identify the components. Later in the Astronomy chapter this will be useful in explaining how elements are detected in stars using spectroscopes.

3. **Show some discharge tubes** (see the Physics Faculty) of some of the elements e.g. hydrogen, helium, mercury.

PART B: Chemical Tests.

These are more complicated and require good self-control by the class members and vigilance by the teacher.

CARE! Acids are corrosive and will harm eyes and sensitive skin even when diluted. 2 Molar hydrochloric acid is used here (as sulfuric acid may complicate issues with sulfides and sulfates). But the teacher may wish to experiment with weaker dilutions. If necessary, some or all of this experiment may be demonstrated and the test-tubes taken around the class for students to smell. Note that the test for sulfides will give toxic hydrogen sulfide gas which has the smell of rotten eggs. If the acid strength used is too dilute, the reaction will not occur. If so, the teacher can demonstrate this reaction by cautiously warming the reaction mixture and then taking the test-tube around the class for a brief time. There seems to be a change in reaction rate when this mix is washed out in water (increases) so have a fume hood handy or a bucket nearly full of water in which the test-tubes can be placed and then carried outside. Good ventilation is needed in the room as even a small amount of H_2S will make some students nauseous.

Also, silver nitrate solution also reacts with chlorides in the skin producing black spots which cannot be erased or washed off. One has to wait until that layer of skin naturally dies and peels off! It will also react with some paints (on walls and desks) and material dyes to give permanent black spots. The students should be cautioned that silver nitrate is toxic so that only one drop should be applied to the chloride solution using a good dropper from a dropper bottle clearly labelled. The reaction with the chloride solution will be a white cloud of silver chloride. This is photosensitive (the silver nitrate and its solution should be in brown bottles) and the teacher can demonstrate this reaction using a beaker of chloride solution and a few drops (say 10) of silver nitrate and then filtering it through filter paper in a filter funnel. This can be placed in sunlight with some object (e.g. an old key) on top. Within a short time, the exposed silver chloride will turn purple (white chloride + black free silver) and leave a white impression of the key.

The appropriate reactions of each of the tests are:

CARBONATES – odourless, colourless carbon dioxide gas is given off as bubbles. The teacher may demonstrate that this is really CO_2 by bubbling some into limewater which turns milky e.g. with calcite:

Calcium Carbonate + Acid = Carbon Dioxide gas + Calcium salt of that acid

CHLORIDES – white precipitate of silver chloride which may turn purple due to black free silver forming with the influence of light breaks up the silver chloride.

e.g. *Sodium Chloride + Silver Nitrate = Silver Chloride + Sodium ions and Nitrate ions.*
Silver Chloride + light = Silver + Chloride ions

SULFATES – a dense white precipitate of barium sulfate forms with the barium nitrate solution.

e.g. *Magnesium Sulfate + Barium Nitrate = Barium Sulfate + Magnesian ions + Nitrate ions*

SULFIDES – toxic fumes of Hydrogen Sulfide gas (rotten egg gas, H_2S) are formed.

e.g. *Lead Sulfide + Hydrochloric Acid = Hydrogen Sulfide gas + Lead Chloride*

QUESTIONS:

1. **Why were each of the specimens finely powdered?**
   To ensure that they will react quickly.

2. **What safety precautions were necessary in this activity?**
   Eye protection, gloves and apron are needed because of the acids and staining chemicals. Use small amounts and restrict movement around the

room. All heating to be very gentle with a low blue flame and the room should be well-ventilated.

3. **What were the special needs of disposing of the wastes? Why? (some class discussion may be needed here!)**
Most chemicals can be flushed down the sink individually with plenty of water but a good classroom discussion on waste disposal is encouraged.

CONCLUSIONS:

1. **List or make a table showing the specimen name, the test performed on it and its result for each metal ion (flame test) and for each non-metal group.**
See above. Students should have these tests recorded in tabular form.

2. **Why was de-ionised water used for making up solutions and dissolving the minerals?**
To prevent any reaction with mineral ions which may be present in drinking water e.g. sulfates and chlorides are common in some water. Teachers may like to test their local water for the class.

3. **What would be the main errors in these tests (be specific for Parts A & B)?**
Main errors involve contamination and the concentrations of acids and solutions. Students are also capable of mixing up the solutions so bottles should be clearly labelled.

4. **In what situations would a scientist in the field use some of the tests in Part B? Explain.**
In the field, specimens are often mixed, are in very small sizes and often masked by weathering products. e.g. gold and pyrite (fool's gold – iron sulfide) are often confused. A test with acid will show which is the sulfide. Similarly, calcite (calcium carbonate) and baryte (barium carbonate) give similar cleavage, colour and rhombic crystals. Whilst baryte has a greater heft than calcite, a flame test (in an ordinary fire) of the powdered minerals will show the green of the barium mineral.

RESEARCH: (Optional)

Use the Internet to find out how geochemists analyse sample using flame spectrometers and other devices.

See:

https://serc.carleton.edu/research_education/geochemsheets/index.html

http://web.uni-plovdiv.bg/plamenpenchev/mag/books/anchem/Handbook%20of%20Analytical%20Techniques,%202%20Volume%20Set.pdf

16.4 Other Activities

- **Form salt deposits** by making a saturated salt solution and then leaving it in a watch glass to allow the water to evaporate. Watch the video from the text:

Online Video: Travel underground into the ancient salt mine at Hallein, Austria.
Go to https://www.youtube.com/watch?v=QCj7CjsYqHw

- **Try panning for gold** using a dirt/crushed pyrite mix. Use gold pans or even wide sided pie tins slightly greased. This activity is best done outside using the pans and a large tub of water. Pan the dirt into the tub. See the demonstration video from the text and watch the video of a reconstructed 19th Century gold town:

Online Video: Watch a demonstration of gold panning technique.
Go to https://www.youtube.com/watch?v=B5h7H4nG7aM

Online Video: Travel back in time to Sovereign Hill, Victoria, Australia - a re-enactment gold rush town, mine and watch a gold smelting operation.
Go to https://www.youtube.com/watch?v=xLQUb9tarNA

- **Contact a local Lapidary Club** or gem dealer/cutter and arrange a visit to their premises or have them come and give a talk about gemstones (with specimens) – another economic resource.

- **Visit a mine site** - there are some problems of Workplace Safety here but worth a try. Alternatively visit the headquarters of a local mining company to talk about methods, uses, equipment, conservation and employment (especially Cadetships and Scholarships).

- Contact the local Resource Council, mining company or government mines department about having a guest speaker come and talk about prospecting, mining and careers in mining.

16.5 Answers to Multichoice Questions

Q1.A Q2.A Q3.C Q4.C Q5.B Q6.B Q7.C Q8.C Q9.C Q10.A

16.6 Some Suggestions for the Review Questions

1. Given two minerals in fine powder form and which each have a gold colour and metallic lustre, how could one quickly determine whether the sample is gold or pyrite (fool's gold)?
   Add a small amount of sulfuric acid to both samples. Gold will not react but pyrite will give the hydrogen sulfide smell of rotten eggs.

2. Discuss the factors which could make a mineral one of economic importance. Could a mineral change its status of being economically important or not?
   It is a matter of usefulness, supply and expense. Minerals and the metals or uses obtained from them can change their status if any of these factors change e.g. iron ores are always in demand because of their usefulness and metals such as gold and silver regularly change their value depending upon market forces.

3. Define each of the following terms:

   (a) **Ore** - a mineral of economic importance
   (b) **Gangue** - unwanted mineral mined with the ore.
   (c) **Stratiform ores** - are ore bodies in layers.
   (d) **Evaporites** - ores formed from evaporation from a solution.

4. Briefly explain, using appropriate mineral/ore examples how each of the following are formed:

   (a) **hydrothermal ore bodies** crystallised in veins or cavities from hot water solutions from a nearby igneous intrusion
   (b) **alluvial deposits** valuable material carried downstream in rivers and on beaches
   (c) **orthomagmatic deposits** valuable ores concentrated by settling and other processes within large igneous intrusions
   (d) **gossans** hard surface cap formed at the top of a vein or other ore body due to weathering of that body at the surface.

5. Research the internet about the use of the metal obtained from haematite:

   (a) in the home
   (b) in the office
   (c) in industry
   (d) in international trade.

Iron from haematite has a great many uses in all of these places. Students could discuss their findings.

6. **Countries such as Australia have huge deposits of uranium minerals which are mined and exported to other countries, yet Australia does not have a nuclear power industry. Use the Internet to review the advantages and disadvantages of nuclear power generation and its possible usefulness in the future.**
Careful with any political discussion here. Many people in Australia are frightened about nuclear power yet overall it has a good safety record. Major disasters in Ukraine, Japan and America have caused some concern. There will be many anti-nuclear websites so one will have to separate the emotions from reality. Personally, the author (who studied nuclear chemistry) believes in a nuclear power industry in the desert on site at Olympic Dam in South Australia but certainly not near major cities. It would be a good interim power source until renewables come up to the level of phased-out fossil fuel. Also, Australia's CSIRO had a good system of waste disposal called SYNROCK.

7. **What is Kimberlite? Why was it so important in the 19th and 20th centuries? Use the internet to find out where other similar bodies are found today.**
Kimberlite is an igneous rock, which weathers to a blue clay which sometimes contains diamonds. It is named after the town of Kimberley in South Africa. Kimberlite occurs in the Earth's crust in vertical structures known as kimberlite pipes, as well as igneous dykes. They occur in many places around the world.

See:
https://www.911metallurgist.com/blog/kimberlite-deposits-and-geology-formation-of-diamonds

8. **What are some of the important products found in evaporative deposits? Name some of the world's most extensive deposits of the minerals.**
Salt (sodium chloride), gypsum, anhydrite, borax, Chile saltpetre, epsomite are just a few of the valuable minerals found as evaporites. They are located in desert regions or areas which were once seas or salt lakes.

See:

https://www.researchgate.net/profile/Hassan_Harraz/publication/301860384_EVAPORITE_SALT_DEPOSITS/links/572a368308aef5d48d30c8a4/EVAPORITE-SALT-DEPOSITS.pdf

9. **Research the source of the metal lithium. What is its importance for the world's future?**
Lithium is an element valuable for the production of glass, aluminium products, and new-age batteries. It is mined as ores such as petalite $(LiAl(Si_2O_5)_2$, lepidolite $K(Li,Al)_3(Al,Si,Rb)_4O_{10}(F,OH)_2$, spodumene $LiAl(SiO_3)_2$ and also subsurface brines. Chile and Australia are the world's largest producers of

lithium with their combined production being greater than 75% of the world's production.

See:

http://www.ga.gov.au/data-pubs/data-and-publications-search/publications/aimr/lithium

10. **Metals such as copper, iron, gold and aluminium are non-renewable resources. Suggest some alternatives to the current excessive use of these metals in industry.**
    This will be a challenge to students as these metals are currently vital to most economies and communities. Alternatives to some of these as jewellery, building supplies and as metal containers could be found but they are vital to electronics and other industries. It may be that the world will have to live with metal production but with improvements in zero-emission power sources and processing and considerable recycling of metals.

16.7 Reading List

Anthony, J. W., Bideaux, R. A., Bladh, K. W. & Nichols, Monte C.,(Edits).(1995). *Handbook of Mineralogy*. Chantilly, VA, US: Mineralogical Society of America. ISBN: 978-0-9622097-1-0

Bullock, L. & Hustrulid, William A. (Edits.). (2001). *Underground Mining Methods: Engineering Fundamentals and International Case Studies*. Littleton, Colo. Society for Mining, Metallurgy, and Exploration. 718 pp. ISBN: 0873351932.

Cox, D. P. & Singer, D. A. (*Editors*). 2014? *Mineral Deposit Models*. USGS Publication. http://pubs.usgs.gov/bul/b1693/html/bull1nzi.htm.

Dixon, C. J. (2012). *Atlas of Economic Mineral Deposits*. New York. Springer Science & Business Media, pp 139. ISBN: 9789401165112

Guilbert, John M. and Park, Jr. Charles F. (1986). *The Geology of Ore Deposits*, New York. W. H. Freeman, pp 985. ISBN: 0-7167-1456-6

Klein, C. & Philpotts, A. (2012). *Earth Materials – Introduction to Mineralogy and Petrology*. Cambridge University Press. 552 pages. ISBN: 9780521145213

Kundu, K. & Kumar, A. (2014). Biochemical Engineering Parameters for Hydrometallurgical Processes: Steps towards a Deeper Understanding. *Journal of Mining: Volume 2014 (2014), Article ID 290275, 10 pages*

Pohl, Walter L. (2011). Economic Geology: Principles and Practice. New York. John Wiley & Sons, pp 680. ISBN: 978-1-4443-3663-4

Chapter 17: Non-renewable Fuels and Energy

17.1 Theme

Human kind has always used fuels to make energy for heating, cooking transportation and, in an expending industrial age, the demand for more energy has become a major world concern. Moreover, the use of fossil fuels has produced problems with air pollution and nuclear fuel has caused some concern with a few major accidents and the problem of nuclear waste.

17.2 Rationale

Of all of the chapters dealing with the usefulness of materials from the Earth and its very environment, this is the most controversial. All societies need fuels for various energy requirements. In large, modern societies, most of the earlier fuels such as wood and peat have been reduced and are used only for small scale energy production. Large scale energy needs are increasing rapidly and the international community has found that the continued use of non-renewable fuels such as fossil fuels is causing damage to the Earth's delicate environment.

17.3 Notes on the Practical Work (see PRACTICAL MANUAL 2.)

Experiment 17.1: Examination of Coal
This is a simple practical exercise observing and describing the various ranks of coal. Use a hand lens or binocular microscope to examine the specimens. It is assumed that class sets can be arranged from coal specimens. Specimens can be obtained from:

http://www.coaleducation.org/resource/sources.htm
(United States)

https://www.amazon.com/Eisco-Bituminous-Specimen-Sedimentary-Approx/dp/B01J4818L8
(world-wide)

https://www.fishersci.com/us/en/catalog/search/products?keyword=anthracite
(world-wide)

QUESTIONS:

1. In some places, peat and brown coal are used in power stations to generate electricity. What would be the disadvantages of using such fuels?
Brown coal has less carbon and water content and so is less efficient. Moreover, it has higher ash so produces more pollution. There may also be more sulfur content so there will be additional sulfur dioxide pollution.

2. What happens to the ash and water when coal is used as a fuel?
Water is driven off as steam and the ash is retained in the furnaces and also some comes off as soot. If it is not captured by filters it goes into the atmosphere.

CONCLUSIONS:

Write a brief description of each specimen giving its name and two key words to remember it's features.

Something like:
> PEAT – woody, fibrous
> LIGNITE - brown, fibrous
> BITUMINOUS – black, banded
> ANTHRACITE – black hard

RESEARCH: (Optional)

1. Use the Internet to find out about peat: how it was formed, how it is mined and where it is still used as fuel.
Peat is only slightly decomposed vegetable matter found in many parts of Europe, America and Australia. It has been traditionally used in Ireland and other parts of Europe as a low-cost but inefficient fuel.

See:
 http://earthresources.vic.gov.au/earth-resources/victorias-earth-resources/sand-stone-and-clay/peat

https://www.britannica.com/technology/peat

2. Use the Internet to research "clean coal technology". How is this done? Are there ways of having a zero emissions coal-fired power station?
It is possible but companies seem reluctant to make use of it. Systems can use algal ponds, or close-system operations to process flue gas from the exhausts of the power stations. Geosequestration, or burial deep below ground in old saline wells is another way to remove carbon dioxide but moisture has to be removed first to prevent corrosion.

See also:

http://www.world-nuclear.org/information-library/energy-and-the-environment/clean-coal-technologies.aspx

https://www.aph.gov.au/About_Parliament/Parliamentary_Departments/Parliamentary_Library/Browse_by_Topic/ClimateChangeold/responses/mitigation/emissions/clean

http://www.rmcmi.org/education/clean-coal-technology#.XE43oGlLfX4

https://www.abc.net.au/news/2017-02-02/clean-coal-explained/8235210

Experiment 17.2: Examination of Some Petroleum Products

A similar experiment to the last except that liquid petroleum products are used. Care should be taken as these are inflammable and will stain skin, clothing and equipment. Samples can be made up from local petrol (gas) stations or hardware stores and kept in smaller glass bottles well-stoppered. Petroleum companies once were a good source of free samples, especially of crude oil but some can be purchased on line. See:

https://publiclab.org/notes/warren/05-03-2013/crude-oil-samples-for-purchase-online

QUESTIONS:

1. What is meant by the viscosity of a liquid?
Viscosity refers to the friction of the moving liquid – the stickiness of the substance.

2. What would happen if the liquid oils in this experiment were all mixed together?
They would mix uniformly.

CONCLUSIONS:

Write a brief description of each specimen giving its name and two key words to remember its features.

Student descriptions should show a range of colours and shades from dark black-green for crude oil to clear kerosene and gases. Viscosity will range from high with crude oil to very low with kerosene.

RESEARCH: (Optional)

1. What are petroleum products used for in addition to being used as fuels?
A great variety of lubricants, manufacture of dyes, cosmetics, plastics and other synthetics See:

http://www.petroleum.co.uk/other-uses-of-petroleum

http://www.petroleum.co.uk/other-uses-of-petroleum

https://whgbetc.com/petro-products.pdf

2. What alternative renewable products could be used instead of petroleum products from crude oil? What are the alternative fuels which could be used instead of petroleum?
Most lists of alternatives refer to alternative fuels and energy sources, but most of the other products such as in the chemical industry could be made from organic oils. See also:

http://www.petroleum.co.uk/alternatives-to-petroleum
(a site for the continued use of oil but a good energy comparison)

https://www.purdue.edu/uns/html4ever/0007.Tao.biofuels.html

Experiment 17.3: Demonstration of the Emission of Fossil Fuel Gases

A good experiment but best done as a demonstration as the equipment is rather elaborate. It could be used as a model for a closed system fossil-fuel power station furnace. The apparatus could be stored away for later use.

QUESTIONS:

1. What was the purpose of the test-tube in ice and containing the silica gel?
This removes any steam/water vapour from the combustion. The blue silica gel will turn pink when saturated with water.

2. Was there any colour change in the Universal Indicator in the flask? What does this show about the gas emitted?
The gas passing through the silica gel and into the universal indicator will be carbon dioxide which will produce an acidic solution of carbonic acid. This will turn the universal indicator yellow to red.

3. Why was the vacuum pump used?
Because there would be insufficient pressure to extract the gases from combustion through the system.

4. If this system was operating using coal or oil as the fuel in a power station, why would the use of such a stage as the ice/silica gel be needed?
Water and carbon dioxide are an acidic mixture which tends to corrode parts and tubing within the exhaust system.

CONCLUSIONS:

1. What gases were given off when the fuel was burned?
Carbon dioxide and water

2. Write a simple word equation for the combustion of the hydrocarbon fuel.

       hydrocarbon + oxygen = carbon dioxide + water

3. How could this system be made into a closed system?
The exhaust gases remaining after the Buchner flask would have to be then passed into additional containers with reagents such as sodium hydroxide etc. to completely remove any remaining carbon dioxide gas.
RESEARCH: (Optional)

1. Use the Internet to research the use of closed systems combustion.
See also notes from the textbook.

http://www.hunwickconsultants.com.au/papers/download/stack_gas_emissions_paper.pdf

https://www.sciencedirect.com/topics/earth-and-planetary-sciences/flue-gas

2. What is geosequestration? How has it been used in the past? What other ways could be used to totally remove flue gas emissions?
Geosequestration is the removal of carbon dioxide from flue gases by pumping it down into deep, natural saltwater wells. It has been used successfully by Norway to remove CO_2 from their North Sea gas wells. The waste CO_2 is pumped back into old wells of the North Sea and below 800 m the CO_2 liquefies and sticks to the pores in the rock.

https://coal21.com/overview/?gclid=EAIaIQobChMI-4HvtpyP4AIVWBePCh2grgnkEAAYASAAEgKSVvD_BwE

https://www.sciencedaily.com/releases/2008/07/080717210554.htm

http://www.ceem.unsw.edu.au/sites/default/files/uploads/publications/solar03_geoseqscenarios_passey.pdf

Experiment 17.4: Nuclear Chain Reaction Data
This is a simple paper exercise based upon nuclear chain reaction data. The final graph should look something like:

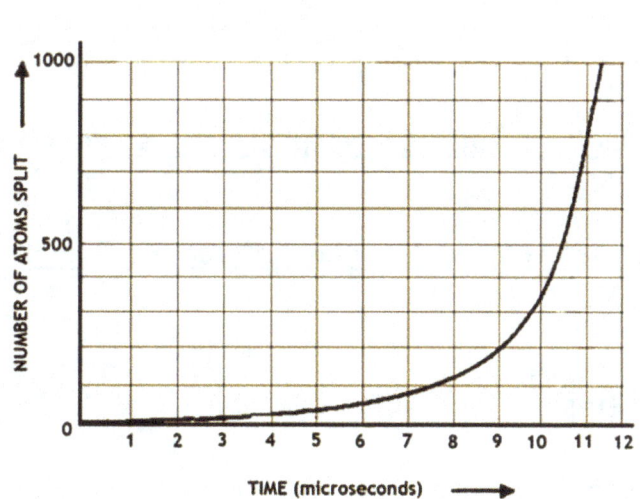

QUESTIONS:

1. What is meant by critical mass?
This is the minimum amount of a given fissile material such as uranium-235 necessary to achieve a self-sustaining fission chain reaction.

2. How is the critical mass attained in a simple fission reactor?
This is obtained by inserting various amounts of fuel rods containing pellets of the uranium fuel into the reactor core.

3. How is the chain reaction controlled in a nuclear reactor?
It is controlled by the use of control rods which contain materials such as cadmium, hafnium or boron, which absorb neutrons from the chain reaction. The moderator, often part of the reactor also absorbs neutrons. The moderator may be graphite rods or heavy water as a pool. See:

http://large.stanford.edu/courses/2011/ph241/grayson1/

CONCLUSIONS:

What is the mathematical relationship between time and the number of atoms split?
It is an exponential function i.e. $y = a\, b^x$

RESEARCH: (Optional)
Use the Internet to locate the closest nuclear reactor. How long has it been in operation?
This will vary from country to country. The only nuclear station in Australia is at Lucas Heights in the south western suburbs of Sydney. It is used to manufacture isotopes and for research. The High Flux Australian Reactor (HIFAR) opened in 1958 and ceased operations in 2007 after being replaced by the Open-pool Australian light water reactor (OPAL), also in Lucas Heights.

The number of nuclear power stations around the world can be found at:

https://www.euronuclear.org/info/encyclopedia/n/nuclear-power-plant-world-wide.htm

https://www.theguardian.com/environment/interactive/2012/mar/08/nuclear-power-plants-world-map

17.4 Other Activities

- Obtain a sample of oil shale and examine its properties or research the Internet about this resource.

- Place some dark oil and water into a measuring cylinder, cover the top and then shake. Watch how the oil and water separate out.

- Make a model of a chain reaction using mousetraps and ping-pong balls or view it:

https://www.youtube.com/watch?v=bc93ig7e4YM

17.5 Answers to Multichoice Questions

Q1.D Q2.C Q3.B Q4.A Q5.A Q6.B Q7.D Q8.A Q9.C Q10.B

17.6 Some Suggestions for the Review Questions

1. Define each of the following terms:

(a) **goaf** – the end of an underground coal mine where the ceiling is allowed to collapse in to be mined.
(b) **gangue** – unwanted minerals mined with ore.
(c) **fissionable** – radioactive isotopes capable of splitting.
(d) **diapir** – dome of salt which wedges up into rock layers.

2. What is coalification? What factors are involved and how does it relate to ranks of coal?
This is the anaerobic process which forms coal from buried plant matter and it depends upon pressure, type of plant matter and temperature.

3. Australia exports coal to China which has a huge coal-mining industry. Use the Internet to explain why this is so.
Australian coal is much cleaner than Chinese coal, having a higher percentage of carbon with less ash, sulfur and water.

4. What is Hilt's Law? Explain how this law affects the ranks of coal.
This states that the deeper the coal, the higher its rank (grade). The law holds true if the thermal gradient is entirely vertical.

5. What factors will determine what methods will be used to mine a coal seam?
The depth of the coal and how the seams are oriented. If the coal is at a shallow depth or if the coal seams are at a steep angle, then open cut may be used. Relatively flat, thick seams at a good depth will be mined underground, especially if there are towns, good farming land or bodies of water on the surface.

6. To what extent has Australian society changed its attitude over time to using alternative energy sources instead of fossil fuels? Justify your answer using examples.
A variety of answers but generally people are becoming more aware of the emissions and costs of fossil fuels, however the alternatives are still limited.

7. Describe the similarities and differences in the geology of coal and oil deposits and how they are extracted from the Earth.

Both coal and oil are formed from the anaerobic decomposition of organic remains. Coal was formed in lacustrine environments from plants and oil was formed in marine environments from marine life.

8. Coal Seam Gas extraction has been involved in much controversy in both the United States and Australia. Use the Internet to list some of the:

    (a) Advantages of Coal Seam Gas - plentiful and relatively easy to extract.
    (b) Disadvantages of Coal Seam Gas - fracking can cause damage to the water table and contaminate water and gas may leak into aquifers.

Warning: Beware of strong bias clouding the real facts from gas companies and action groups.
Note: there is some difference between the US experience and that in Australia.

https://www.aph.gov.au/About_Parliament/Parliamentary_Departments/Parliamentary_Library/pubs/BriefingBook44p/GasDebate

https://www.business.qld.gov.au/industries/mining-energy-water/resources/petroleum-energy/outlook-statistics/petroleum-gas

https://theconversation.com/coal-seam-gas-emissions-lower-than-us-first-australian-study-29699

9. Research the alternatives to the use of fossil fuels with special reference to their initial set-up costs, long term costs and expected lifetime and disposal.
Initially the use of alternative energies looks bright but one must consider the initial costs and the use of non-renewable resources in building alternative energy systems. If the short-term costs are high but the log-term use is very long then the use of the alternative will be cost effective. See:

https://www.aph.gov.au/About_Parliament/Parliamentary_Departments/Parliamentary_Library/Browse_by_Topic/ClimateChangeold/responses/mitigation/emissions/alternative

10. What is meant by the following terms and what is the current state of their development and application?

(This is a difficult question which needs to be answered - probably as a group discussion with one topic per group after some Internet and other research):

(a) Geosequestration of carbon dioxide - burying the dry gas in old saline wells deep underground. Not well-developed, especially in Australia.

(b) Flexible, thin-film solar panels - still in its infancy with most solar panels being of the silicon rigid type. See:

    https://www.solarquotes.com.au/panels/flexible/

(c) Photobioreactors and algae – use of algal ponds or racks of containers to absorb flue gases. Still at the experimental stage.

(d) Clean waste-removal processes – using high temperature furnaces is on-going in many countries but yet to fully replace waste dumps.

(e) Clean coal technology – possible but costly and is a matter of scale. Not yet developed fully.

17.7 Reading List

Elliott, David. (2007). *Nuclear or Not? Does Nuclear Power Have a Place in a Sustainable Energy Future?* London: Palgrave Macmillan. ISBN 978-0-230-50764-7.

Freund, P. and Kaarstad, O. (2007). *Keeping the lights, Fossil Fuels in the Century of Climate Change.* Oslo, Norway: Universitetsforlaget. ISBN 978-82-15-01141-7

National Oceanic and Atmospheric Administration (NOAA). http://www.noaa.gov/

Simon, Christopher A. (2006). *Alternative Energy: Political, Economic, and Social Feasibility.* Lanham, Maryland: Rowman & Littlefield, 2006. ISBN 0-7425-4909-7.

Sovacool, Benjamin K. (2011) *Contesting the Future of Nuclear Power: A Critical Global Assessment of Atomic Energy.* Singapore: World Scientific. ISBN: 978-981-4322-75-1

Tester, Jefferson W., et al. (2005). *Sustainable Energy: Choosing Among Options.* Cambridge, Massachusetts: The MIT Press. ISBN 0-262-20153-4.

The Intergovernmental Panel on Climate Change Fifth assessment report. http://www.ipcc.ch/

Zehner, O. (2012). *Green Illusions: The Dirty Secrets of Clean Energy and the Future of Environmentalism.* Lincoln, Nebraska: University of Nebraska Press ISBN 978-0-8032-3775-9.

Chapter 18: Exploration for Resources

18.1 Theme

Exploration scientists are trained to observe the landscape and the processes of the Earth systems very carefully. They look beyond what the untrained person sees, especially looking for patterns or differences occurring in the natural world. To assist them, there are a number of techniques and instruments which can be used to detect or measure some of these differences or natural qualities.

18.2 Rationale

It is important to know how the field scientist works and some of the techniques and instruments at their disposal operate. This chapter gives but a brief view of some of these techniques and instruments but full competence requires professional experience and practice.

18.3 Notes on the Practical Work (see PRACTICAL MANUAL 2.)

Experiment 18.1: Simulated Aeromagnetic Survey

This experiment attempts to show how an aeromagnetic survey would work by using a very simple model. Aeromagnetic surveys are an excellent way of searching for minerals and structures which have some effect on the Earth's natural magnetism. A very large area of surface can be covered using an aircraft or helicopter towing an air-borne magnetometer which will give a continuous readout from the instrument. Navigation is important so the aircraft will fly along a designated pattern relative to the ground. This enables ground surveys by field scientists who would walk along these patterns and do a much more detailed search.

Practice has found that a piece of magnetic strip obtained from old refrigerator doors works well but any low-powered magnet, such as a fridge magnet which is used to attach messages to a metal door, will do. The important thing is that the magnet must not be too strong otherwise its field completely dominates the area and compass needles are too easily attracted. Small compasses are also good for the same reason. They are small enough to act as simulated magnetometers over the area of the grid. They should have a readable compass rose measured in degrees. To build up a useful grid of values equivalent to a fine aeromagnetic survey, the halfway points between each grid intersection are also given a value found by calculating mathematically the average value between each grid value. Isogonal lines, those lines joining places of equal magnetic value can be drawn (for say, whole value multiples of ten etc.) around the map by estimating where each value would be between the numbers on the grid.

QUESTIONS:

1. What is meant by a magnetic anomaly?
A difference in what the Earth's magnetic field should be at a given locality.

2. What assumption is made by finding the averages between grid values?
That the magnetic field between these points is uniform.

3. When the compass is directly over the ore the readings may be zero. Why?
Because the magnetic field is strongest and acting directly down below the compass. Ideally a magnetic dip meter showing the dip or angle from the horizontal would be a better meter.

CONCLUSIONS:

What is the map location of the suspected ore body? (Use the alpha-numerical labels e.g. A 1 etc.)
This will be determined by where the magnet has been located. Students could play a game of hide-and-seek with one student locating the magnet and the others doing the aeromagnetic survey. Also, a piece of magnetite could be tried as a substitute for the magnet.

RESEARCH: (Optional)

Other than searching for ore bodies, what are the other uses of aeromagnetic and ground magnetic surveys?
They can also be used as a wide-area mapping of the geology of the sub-surface of a region, especially when looking for potential oil traps, faults and ancient rock types.

See:
http://www.cas.usf.edu/~cconnor/pot_fields_lectures/Lecture8_magnetics.pdf

https://www.geosoft.com/media/uploads/resources/technical-papers/Aeromagnetic_Survey_Reeves.pdf

https://pubs.usgs.gov/bul/1924/report.pdf

Experiment 18.2: Introduction to Aeromagnetic Survey Maps

This experiment is a computer exercise which follows on from the previous experiment. It shows a map of California USA in false colours generated by a computer to show magnetic anomalies in nanoteslas. The aim of this experiment is to give students a chance of interpreting such a map to locate geological structures, especially faults and basins. They will be able to compare the magnetic anomaly map with a geological map of the same area and thus make meaningful comparisons. The lesson could include a final group discussion about the colours on the magnetic map and the corresponding geological features. Observation is encouraged.

1. Magnetic anomaly methodology and interpretation is more complex than what is inferred in this simple exercise. What other factors must be considered when looking at such a map?

Weathering and erosion will change some of the magnetic properties of the surface.

2. What are some possible errors which could occur in the making of such a magnetic anomaly map (some extra Internet research may be needed on aeromagnetic surveys?

General magnetic interference due to man-made structures such as powerlines, metal structures and mine workings.

CONCLUSIONS:

Write a general conclusion on the use of aeromagnetic surveys and some of the general matches which have been found between the false colours, shapes and the geological setting.

Students could pool their results to obtain a basic set of principles about reading magnetic anomaly maps.

RESEARCH: (Optional)

Research the lives and work of Nikola Tesla and Wilhelm Weber

See:

https://theconversation.com/nikola-tesla-the-extraordinary-life-of-a-modern-prometheus-89479

https://www.smithsonianmag.com/history/the-rise-and-fall-of-nikola-tesla-and-his-tower-11074324/

https://nationalmaglab.org/education/magnet-academy/history-of-electricity-magnetism/pioneers/wilhelm-weber

http://www-groups.dcs.st-and.ac.uk/history/Biographies/Weber.html

Experiment 18.3: Interpretation of Magnetic and Gravity Data Sets

This experiment extends the previous two experiments by showing students that field scientists do not rely only on one set of remote data but will use several before embarking on ground surveys. They recognise the limitations of separate sets of data and so that they overcome any remote errors by using and comparing several data sets. This is a very general comparison over an entire continent so only large-scale general features and geological structures are going to be found. Group discussion is probably useful here to make up a set of comparisons for general interpretation. It also gives students some general overview of the nation's geological structure.

QUESTIONS:

1. What does Map A show about the island of Tasmania (bottom of map)?
That it is really part of the mainland – geological structures continue under Bass Strait.

2. Which of the two anomaly maps seemed to show more detail?
This question refers to Maps A and B. They both show different features but Map A seems to be complicated in the features shown.

3. What units are used in measurements on the gravitational anomaly map?
Map A uses nanoteslas (nT), a unit of flux or field strength per area, whilst Map B measures in micrometres per second squared (μ m s^{-2}) which is an acceleration unit caused by the gravitational field i.e. comparing the values to that of normal gravitational acceleration 9.8 m s^{-2} so the units on the map are very small differences.

4. Why are Bouguer anomaly maps adjusted to terrain and height?
Because the gravitational field decreases with distance from the Earth's centre and therefore its surface above sea level i.e. the units need to reflect what is under the surface not how far the surface is away from the centre of the Earth gravitational pull.

CONCLUSIONS:

Write a general conclusion on the usefulness of geophysical survey maps such as magnetic and gravity anomaly maps referring to some of the general matches which have been found between the false colours, shapes, geological structures and the surface geology.

Again this would be a useful student group discussion in deriving magnetic colours and shapes to real geological structures e.g. one should see such structures (especially on Map A) as the Yilgarn Shield and Darling Fault line of Western Australia (on the lower left of the Shield, near the coast), the wide Murray-Darling Basin of central New South Wales and the New England Basin on the border of NSW and Queensland.

RESEARCH: (Optional)

Research the life and work of Pierre Bouguer
See:

http://www-history.mcs.st-andrews.ac.uk/Biographies/Bouguer.html

https://www.revolvy.com/page/Pierre-Bouguer

Also see: Explanation of magnetic surveys at: https://youtu.be/AZyNIGFHsE4

Explanation of gravity surveys at: https://youtu.be/9P6GEpxFtSY

Experiment 18.4: Interpretation of Radiometric Data Sets

Radiometrics is another type of data set useful to the field scientist, especially those interested in the nuclear radiation levels of the surface and sub-surface. The data can be found using aerial surveys and then supported using a detailed ground survey. Usually the field scientist is interested in natural emissions of radioactivity which could indicate a source of radioactive ore such as those for uranium, thorium and others. The experiment follows the same method of the previous experiment where students are asked to compare a radiometric aerial survey with a national geological map. The notes and maps of specific radioactive elements are given at the right of Map A and the notes in particular are good to read and summarize. These maps are easier to follow and give some clear localities of concentrations of these elements. The white areas are places yet to be surveyed, showing that there still is a lot of the Earth that is still unknown. Each of the smaller maps have good explanations of what is to be found, but in general, red = a high concentration of the element and blue = a low concentration.

QUESTIONS:

1. What type of rock often contains high concentrations of uranium?
Usually deep crystalline igneous rocks such as granite. Look at the small map for uranium and one can clearly see the Snowy Mountains granites at bottom right and the Granite Belt on the eastern New South Wales – Queensland border.

2. What do the white parts of the map represent? What does this suggest?
As discussed above, these are areas yet to be surveyed and show that there is still much exploration to be done.

3. Australia has some of the biggest uranium reserves in the world and actively exports uranium. Where are the major mines located?
See:
http://www.world-nuclear.org/information-library/country-profiles/countries-af/australia.aspx
(scroll down to the map)
The underground mine at Olympic Dam, South Australia is the largest uranium mine in the world (and the fourth largest copper mine).

CONCLUSIONS:

Write a general conclusion on the usefulness of radiometric survey maps referring to some of the general matches which have been found between the false colours, shapes, geological structures and the surface geology.
This should come from the student discussion of their results.

RESEARCH: (Optional)

1. What are some of the problems associated with uranium mining in Australia?
2. Prepare a short talk or fact sheet for the case (a) for and (b) against the development of nuclear power in Australia.

Both these topics are controversial as Australia has the biggest supply of nuclear fuel but no nuclear power industry. The country relies on fossil fuels for most of its electrical supply. The lack of nuclear power stations reflects an emotional social attitude rather than the lack of knowledge and ability of engineers, scientists and managers to construct and operate nuclear power stations. Personally, the author advocates the use of nuclear power in Australia to meet the shortfall between phasing out fossil fuel power stations and large-scale alternative energy use. The main objections concern the proximity of power stations to habitation, nuclear accidents, transportation of nuclear fuel and wastes through habituated areas and the removal of nuclear wastes. These problems can be overcome by:

- Building the power stations on site or near Olympic Dam and other remote sources of Uranium (look at the map). Private research suggest that this is possible even though transmission of power will require building transmission towers across long distances which will be initially expensive but cost effective in the long term.

- Nuclear accidents are uncommon and the industry is considered safer than that of the fossil fuel industry. Unfortunately, when an accident happens it is a MAJOR disaster. An analysis of the world's great nuclear disasters will show that they have occurred through faulty maintenance and safety or being built in earthquake and tsunami zones. Many of the nuclear stations in California are built on fault lines but all of these accidents can be avoided with planning and care. Australia is one of the most stable continents on the planet and a power station in the desert hundreds of kilometres from habitation ensures a good chance of stability.

- If the station is built on site or nearby and connected by rail then there would be no problem about the transportation of nuclear fuel through centres of habitation. Similarly, with nuclear waste which, like the fuel is low in volume compared to fossil fuel transportation.

- Removal of nuclear waste has been a problem in Europe and America where there are high concentrations of Human habitation. In the days before the problem became critical, nuclear waste was stored in drums in old mines, or simply dumped out at sea or covered over by cement on land. The latter technique was used in the South Australian desert following the British nuclear-testing program in the 1950s and 1960s. These drums have been subject to erosion and exposure. Synroc (synthetic rock) is an artificial material created by a joint program between ANSTO, the Australian Nuclear Science and Technology Organisation and the Australian National University. It is designed to be formed with and around spent nuclear fuel to form a solid, water-proof material which could then be buried deep below the Earth in drill holes drilled into the many hard, crystalline rocks of ancient cratons well away from human habitation.

Experiment 18.5: Interpretation of a Geochemical Data Set for Gold

This is another data set computer exercise in which the students can use geochemical data and make comparisons with the nation's geology. As before, it is important that the students read the information which comes with the main map and then make collective judgements about the data set interpretations. They will have to scroll down to the maps on the first site. The second site has some good data about locations of gold mining and exploration as well as data about gold in Australia - one of the world's most productive gold centres.

QUESTIONS:

1. How were the samples collected? (refer to number of sample sites and the meanings of TOS and BOS).
Samples were collected from sediments with TOS referring to Top Outlet Sediment (0-10 cm depth) and BOS being Bottom Outlet Sediment (60-80 cm depth, on average).

2. How were the samples prepared and analysed? What precautions were taken to prevent errors?
Most samples were taken from near the mouth of any catchment with trial testing in some known areas. Many samples were composite soils to reflect the local soil identity. At each locality a detailed site description, field pH, and dry (if possible) and moist soil colours were recorded and several digital photographs were taken. All samples were sent to Geoscience Australia (Canberra) for processing, where they were air-dried, homogenized dry sieved. Sample numbers were randomized to minimize regional bias, help separate false from true anomalies and obtain meaningful estimates of the variance of duplicates. Field duplicates, analytical duplicates, internal standards and certified reference materials were introduced at regular intervals in the analytical streams. Care was also taken throughout the project to minimize contamination, cross-contamination and mis-labelling.

3. Comparing gold to other metals, what is a unique property which would aid in its sampling?
Gold is chemically inert so it has not reacted with the environment to form complex compound ores.

4. Is there a match between gold mining locations and this new sampling data set? Give some locations which have a good match.
Scroll down to the big map on Page 19 and look for high concentration of gold indicated by squares. Compare this to the gold-mining localities of the second site. There is a general match between the squares and major gold-mining areas, especially central New South Wales and north central Queensland.

5. Are there any locations on the National Geochemical Survey map which are not currently mined? Comment on their potential.
Looking at both maps and especially the higher concentrations, there are several interesting sites such as near the Queensland, South Australian and Northern Territory borders and south of Batman in Northern Territory.

CONCLUSIONS:

1. Write a general summary of the methods used in the national survey.
2. Write a general conclusion on the usefulness of geochemical survey maps referring to some of the general matches which have been found between the survey and known gold mining locations.

These questions should come from a group discussion of the experiment and the answers to the questions above. The experiment should show that such an elaborate geochemical data base is very useful in locating potential sources of gold.

RESEARCH: (Optional)

Use the Internet to find out about how gold is used.

Gold has a great range of uses apart from jewellery and electrical conductivity in electronic devises.

See:

https://geology.com/minerals/gold/uses-of-gold.shtml

http://goldresource.net/modern-uses-of-gold/

Experiment 18.6: Interpretation of Drill Hole Logging

This activity is an exercise in drill hole logging one of the basic tools and data sets common to many geological explorations. Students who go on to work in the industry will encounter drill cores as part of their investigations. Drill cores are obtained from drilling rock with hollow tubes with a cutting bit at one end. As the stick is drilled into the ground, more can be screwed on to the ends so that the drill may go down hundreds of metres. These can then be hauled to the surface, opened and the drill cores, as cylinders of rock about 5 cm across, are then laid out in order in metal trays which are then stacked in drill-core libraries as samples of that drilling operation. Many companies have their own drill-core libraries but there are government libraries which offer access to their cores. Cores are logged very much like an outside cliff-face with distance and rock type and structure being noted and drawn like a stratigraphic column to be interpreted and matched in the same way.

In addition, various devices can be lowered down the drill hole to measure a variety of parameters such as electrical resistivity, changes in electrical voltage (SP), radioactivity and many other parameters. The experiment is fairly self-explanatory.

RESULTS:

1. Draw the stratigraphic column to scale. Use conventional symbols and colours to represent the lithologies and show these as a legend.

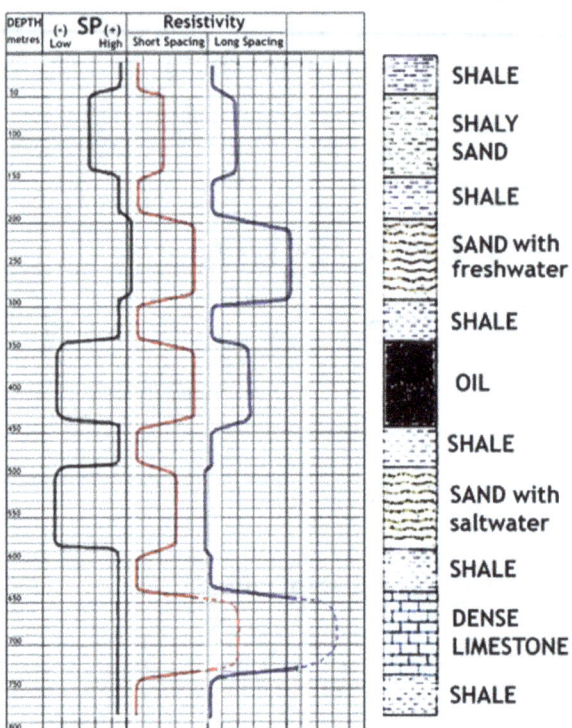

2. Give the depth to the (a) freshwater aquifer and (b) the oil trap.
200 metres to the top of the aquifer and 500 metres to the top of the oil trap.

QUESTIONS:
1. What is the relationship between electrical resistivity and conductivity? What is the unit of resistivity?
Resistivity, the ability of a material to slow electrical current is the inverse of conductivity which allows it to flow in materials (conductors, semi-conductors). Resistivity can be measured in ohms-metre (the SI unit) or Siemens per metre.

2. What factors determine the resistivity of a rock or sediment?
Factors include the nature of the rock or soil, its moisture content and any ions within the water, its density and temperature. See:

https://orkustofnun.is/gogn/unu-gtp-sc/UNU-GTP-SC-11-21.pdf

3. Why would crude oil have lower resistivity than freshwater? (some research needed)
This is probably an unfair and theoretical question. Whilst hydrocarbons usually have high resistivity, they may still contain some charged ions whereas completely

freshwater (uncommon) will have few if any ions and so have higher resistivity. Most freshwater encountered in wells will have some ions and so lower resistivity.

4. Why would the long spacing resistivity results be more reliable than the short spacing?

This refers to the distance of separation of the electrodes. Longer spacing will measure SP with less contamination from the lubricating drill mud nearest (short spacing) the drill.

5. Why is the curve for shale often used as a standard Shale Line for comparisons?

The Shale Line marks the resistivity for shale rocks and is used as a comparison to other resistivities because shales are impermeable and will not contain fluids so have relatively constant values of SP and assumed to be zero. See:

https://www.uio.no/studier/emner/matnat/geofag/GEO4250/v08/undervisningsmateriale/Lectures/BWLA%20-%20Spontaneous%20Potential%20-%20Gamma%20Ray.pdf

CONCLUSIONS:

1. Write a general summary of the usefulness of e-logs in exploration geology.

This could come from student discussion and opinion, but generally such logs are extremely valuable as they and their cores can give a great variety and amount of data from one drilling.

2. Write a specific evaluation (as an exploration geologist) of the geological setting and its potential for exploitation, giving details of the lithology, any resources found and any other factors about their extraction.

From the log it can be seen that there is both potential for freshwater and hydrocarbon resources. The geological setting is marine.

3. What are some of the errors which may give inaccurate results in using this method? How could they be compensated?

Errors involve the type of drill mud used and the type of data instrument which will be sent down the hole e.g. for resistivity, a hydrocarbon mud reduces error because there is less electrical interaction between the mud and the electrodes. Depth estimation can also introduce an error in the logging of the rocks as can the expertise of the person doing the logging on the surface. Other factors include the geological setting such as striking faults or harder rock.

RESEARCH: (Optional)

Use the Internet to find out about other types of electric logs which could be used, especially those which may reinforce the findings of this SP/resistivity log.

There are many different types which can be chosen by the geologist to match the potential geological setting. These include radiometric devise, acoustic logs, nuclear magnetic resonance and many others.
See: https://petrowiki.org/Types_of_logs

18.4 Other Activities

- Obtain a box of drill cores if possible, from a mining or oil company drill core library. They sometimes discard old cores. Also see government mining departments to see if they can donate cores. They usually come in a metal box of about a metre by 30 cm by about 10 cm. This gives about 5 metres of drill core.

- Visit a local drill core library (also see government agencies and local mining companies).

- Invite a drilling team to visit – they may have their drill truck with equipment. They can talk about drill cores or drilling for water.

18.5 Answers to Multichoice Questions

Q1.A Q2.C Q3.B Q4.A Q5.B Q6.A Q7.D Q8.A Q9.D Q10.B

18.6 Some Suggestions for the Review Questions

1. **What resources would be researched before applying for a mining lease?**
   This will include legal and traditional ownership of the surface of the land, environment including any restrictions concerning endangered species, access roads and local infrastructure, climate and water resources and many more (see text).

2. **Suggest some of the problems which would be encountered when applying for a mining lease to mine a shallow seam of coal in an area which is ten kilometres from a town, thirty kilometres from the sea and in a known region of previous indigenous habitation. How could these problems be overcome?**
   Some major problems which need to be overcome. This scenario is typical of what happens in many areas. The seam is shallow so it will be mined by open pit which will require access to the site and considerable disruption to local and traditional owners, perhaps sacred sites and local farms. Natural environments will also be destroyed so an environmental impact statement and appropriate environmental preparation, collecting and husbandry will be needed. Whilst there is close proximity to the coast for the potential of building export facilities, there might also be the potential for water pollution from the site. Townspeople may welcome the jobs in the region but there also will be problem with excessive road transport and dust.

3. **List all of the equipment that would probably be needed for a two-day prospecting walk up a stream in a temperate climate looking for gold.**

   A good class discussion and if close to a designated gold reserve (see local tourist bureaux), a good planning phase for a day or weekend trip. Equipment will consist of that required for the gold panning (pans, sieves, shovels), camping (shelter, food, water, utensils), clothing appropriate to the expected climate and transportation. Safety issues are also an important factor and if going on the excursion a full risk assessment and communication/safety times must be considered.

4. **GPS units have become a very useful tool for field scientists. Give at least two uses and problems:**

   (a) **practical uses for having a hand-held GPS unit in the field** – very useful for general location recording and also for navigating to places with known coordinates.

   (b) **instances when a GPS unit is likely to fail** – they tend to fail unexpectedly and when needed due to lack of signal, batteries uncharged (solar power supplies available), extreme weather (heat, cold, wet).

5. **As an environmental scientist working for a mining company, what advice would you give to the company before it begins work on a proposed open-cut gold mine?**

   See the notes above about environmental considerations. Mining companies hire environmentalists to avoid problems. Advice should be well-researched such as the general environmental setting, endangered species, needs for collection, keeping and restoring plants, escape routes for animals and strategies for their return after rehabilitation, water resources and potential pollution problems with water, air and soil etc.

6. **Use the Internet links to summarize the rights of landowners in regard to the minerals on their land.**

   Many of these sites are given in the text book and include State and Federal Government acts.

7. **What instruments could be used by a geophysicist to measure the Earth's local:**

   (a) **magnetic field** – airborne or ground magnetometers.

   (b) **gravitational field** – airborne or ground gravimeters.

   (c) **Radioactivity** – airborne or ground scintillation counters.

   (d) **geothermal heat flow** – airborne infra-red images, ground temperature logs (drill) and digital temperature probes.

   (e) **rock porosity** – drill cores and logs of resistivity.

8. Compare and contrast:

    (a) **push broom scanning and whisk broom scanning** – these concern the way that various scanners on board satellites gather the data. Push broom scanners sense data using a line of scanners which move at right-angles to the direction of the satellite's path. Whisk broom scanners also move across the path direction but they use a mirror to sent the light to a single detector.

    (b) **hyperspectral and multispectral scanning as satellite scanning methods** - multispectral imagery generally uses a limited number (3 to 10) of bands of electromagnetic radiation using a remote sensing radiometer whereas hyperspectral imagery has a large number of narrower bands (10-20 nm) which are detected and also analyzed by an imaging spectrometer.

9. **Use the internet to find out what geophysical equipment is used to target:**

    (a) **Gold** – e.g. ground penetrating radar, electromagnetic conductivity equipment.
    (b) **metal sulfides** – e.g. gravitational surveys.
    (c) **Uranium** – e.g. radiometric surveys.

10. **Research the advantages and disadvantages of being a field researcher requiring at least a month in the field working from a base camp in a remote area. How will the nature of the land surface (topography) and climate affect the field work?**

    A good discussion about living in the field for extended periods. Most students have little knowledge of such an experience. Apart from the scientific equipment and methods needed, they should consider living conditions, sanitation, food/water supplies, transportation and the need for extra supplies etc. All of this will vary with the terrain which will restrict movement, the climate and the environment including dangerous plants, animals and geological dangers.

18.7 Reading List

Burger, H. Robert; Sheehan, Anne F.; Jones, Craig H. (2006). *Introduction to Applied Geophysics: Exploring the Shallow Subsurface*. New York: W.W. Norton. ISBN 0-393-92637-0.

Coe, A. L. (Edit.) (2010). *Geological Field Techniques*. London: Wiley-Blackwell. ISBN: 9781444330625

Conklin, A. R. Jr. (2004). *Field Sampling: Principles and Practices in Environmental Analysis*. London: CRC Press - Taylor & Francis Group. ISBN 9780824754716.

Faure, G. (1998). *Principles and Applications of Geochemistry: A Comprehensive Textbook for Geology Students*. Upper Saddle River, NJ: Prentice-Hall. ISBN 978-0-02-336450-1.

Kearey, P., Klepeis, K.A., Vine, F.J. (2013). *Global Tectonics*. London: John Wiley & Sons. ISBN 978-1118688083.

Knödel, K, Lange, G. & Voigt, H. (Eds.). (2007). *Environmental Geology - Handbook of Field Methods and Case Studies*. Berlin: Springer-Verlag. ISBN 978-3-540-74669-0

McCoy, R. (2005). *Field Methods in Remote Sensing*. New York: The Guilford Press. ISBN 1-59385-080-8.

Watts, S. & & Halliwell, L. (1996). *Essential Environmental Science: Methods & Techniques*. London & New York: Routledge. ISBN: 0415132479

Wheater, C. P., Bell, J.R. & Cook, P.A. (2011). *Practical Field Ecology: A Project Guide*. London: John Wiley & Sons. ISBN: 978-0-470-69429-9

Chapter 19: Mining Economic Minerals

19.1 Theme
Economic ores are the source of the metals and other useful resources used in modern communities. It is difficult to image a world without these resources yet they are non-renewable and mines only have a limited life before all of their ores have been extracted.

19.2 Rationale
An understanding of mining techniques and how they are used will enable a deeper knowledge of the need for these resources and continued development of mining which has minimal impact on the environment. In the past, mining proceeded without much care of the environment and considerable damage was done. To some extent this still goes on in some countries where the environmental laws are relaxed. Responsible mining ensures that the environment will be protected and rehabilitated after the mine has closed. This is one area where Earth and Environment Scientists can be employed and play a major role in restoring the planet.

19.3 Notes on the Practical Work (see PRACTICAL MANUAL 2.)

Experiment 19.1: Virtual Excursion - Open Cut Mining

It is very difficult to organize an excursion to a mine site, even when mining companies sometimes welcome visitors. Health and Workplace Safety issues often do not allow students on to sites unless there has been considerable planning on both sides. These virtual excursions are an attempt to give some idea of the mining operation with student control on how the videos are viewed and studied.

Part A: Cadia Valley Gold and Copper Mine, New South Wales, Australia.
This video was made by the author during a student visit sponsored by the Australian Institute of Mining and Metallurgy. Open Pit mines usually cover large areas either because the resource such as coal is in extensive flat, nearly horizontal seam or because the metallic resource is spread throughout the country rock.

https://www.youtube.com/watch?v=ohxxYSjcXa4

The operators of the Cadia mine are supportive of the local community and the environment and pride themselves in the efficiency of their operation which uses radio coordination of each phase of the mining process. See:

http://www.newcrest.com.au/our-business/operations/cadia-nsw/

Part B: Gold Mine, Nevada, USA – an extensive virtual excursion.
Sometimes the resource is mined to the limit of Open Pit mining and so an underground mine is started at the bottom of the pit. This excursion visits a much larger mine and offers a much more detailed look at the operation. Take the time and opportunity to explore this site fully with special reference to the life and work of the people engaged in the operation.
Open the site at:

http://clickschooling.com/2017/10/nevada-gold-mine-virtual-tour/

This website allows access to an interactive tour of the mine and its support operations at:

http://xplorit.com/nevada-mining-web#

Part C: Black Thunder Open Pit Coal Mine
In contrast to gold mining, this mine extracts coal using similar techniques. It is in the U.S. state of Wyoming, located in the Powder River Basin and contains one of the largest deposits of coal in the world. In large open pit coal mines, much of the same operation occurs but the coal is usually sieved on large screens and washed before the final storage phase.

Open the site of the video at:

http://www.rmcmi.org/news/detail/2011/05/03/take-a-virtual-tour-of-black-thunder-coalmine#.XC2T4WILfX4

There has been some difficulty in opening this website so try going to:

https://www.youtube.com/watch?v=2LQwxTm94Ps&feature=share

Also see their site (with chat facilities) at:
http://installatiewerkenutrecht.nl/coal-9914-Black-Thunder-Coal-Mine-Website/

QUESTIONS:

1. In general, and from the textbook notes and the videos, what would be the sequence of events before the mining company arrives on its site?
See the steps given in the textbook.

2. What happens next after the environmental phase when seeds and seedlings are collected from the area and escape corridors are made to allow animals to leave the area?
The topsoil, a valuable resource for future re-planting, is removed and stockpiled so that it can later cover the back-filled pit.

3. Having exposed the rock surface, how is the mining process started?
The pit is started, usually at a productive end by drilling shot holes so that the rock can be blasted out to get to the ore.

4. Why are the haulage roads made in a spiral manner around the outside of the pit? What would be some of the safety considerations and strategies used to ensure the safety of truck drivers and the stability and surface of the roads?

This enables a good low-angle grade capable of carrying trucks in safety. These roads are engineered so that they are not in areas where they will give way because of fractures in the country rock. Moreover, their sides are cut at angles which limit collapse. They are constantly cleared, graded and hosed down to limit dust.

5. How is the material exacted and transported? Where is it taken?

The rock without ore is dumped into trucks with large excavators and the rock taken to stockpiles to later fill in used parts of the pit. Rock with ore is loaded in the same manner and taken to the crushing plant to be crushed into fine particles and processed.

CONCLUSIONS:

Summarize the main operations observed in the open cut process. A table giving the various phases of operation for each of the three locations may be useful.

This is to be done from student notes made after the videos have been watched. Repeat showings may be needed with a class discussion to summarize the main points.

RESEARCH: (Optional)

Use the Internet to find out about the locations of each of these mines and the minerals mined.

The general names and locations as well as websites are given, Google Maps can be used as a whole class activity to locate these mines.

Experiment 19.2: Virtual Excursion – Deep Shaft Mining

In contrast, deep shaft mining requires different techniques and a different breed of miner used to working in confined spaces, often with equipment and toxic fumes. Deep shaft mining is a culture all of its own. Often the mining is done many hundreds and sometimes thousands of metres below the surface.

Part A: Central Deborah Mine Gold Mine, Bendigo, Victoria, Australia.

This video was made by the author during a student visit to this deep gold mine, the upper levels of which are often open to visitors but mining and blasting continued below. Open the site of the video at:

https://www.youtube.com/watch?v=k24OTTg9F60

Part B: Mponeng gold mine in South Africa's Gauteng province - the world's deepest mine.

This is a commercial documentary lasting about 30 minutes. It may be shown in sections if time does not permit the full showing. Open the site of the video at:

https://www.youtube.com/watch?v=6ZtYlnuOKtE

QUESTIONS:

1. In what circumstances are deep shaft mines used?
When the ore body is inclined or follows a twisting direction and or when there are surface features such as good farmland and towns to be kept intact.

2. What type of equipment is used in deep shaft mining? Compare this type of mining with that of open pit.
This varies with the mine. In some deep shaft mines equipment may be compact, even remote-controlled but they also could be of a normal size, constructed in workshops in the mine and then left there when the mine is finished. Open pit mines allow for larger equipment with huge excavators, haulage vehicles and other large plant equipment.

3. What are some of the major problems associated with deep shaft mining concerning:

 (a) construction – they follow the ore so a variety of construction methods are needed in building ceilings and walls against collapse, often using mesh and ring bolts to hold the ceiling.

 (b) prevention of collapse – mesh and ring bolts with some supporting walls and beams in places.

 (c) working conditions – often cramped and sometimes requires cooled forced ventilation. Lunch (crib) rooms and rest rooms are built in the mine.

 (d) safety – concerns collapse of the mine and the proximity of machinery such as excavation equipment, crushing plants and conveyor belts. Often strict rules apply but accidents, mainly collapses, do happen.

4. Why are deep shaft mines hot? Research how they are cooled to help working conditions.
The Earth becomes hotter with depth at about 25-30 °C/km. Forced air circulation is required with the injection of liquid air (cryogenic chilling) at -190^{0C} sometimes used in deeper and hotter mines.

5. How is the material exacted and transported?
In deep mines, the ore is often crushed in the mine and sent above by conveyor belts. In some mines it may be loaded onto flattened trucks to be hauled up long spiral access ramps.

CONCLUSIONS:

Summarize the main operations observed in the deep shaft mining process. In the summary include the reasons for such mining, the methods used in excavating the mine, any engineering processes designed to hold the mine together, how material is excavated and removed and the main problems associated with such mining.
This could be done from a group discussion.

RESEARCH: (Optional)

Use the Internet to find out about the locations of deep shaft mines in the local area, what resource they extract and how this resource is transported and marketed. Also research the dangers of abandoned mine shafts in the local area or in general.
If there are no mining facilities or old workings nearby use the nearest area or choose one within the state.

Experiment 19.3: Virtual Excursion - Deep Shaft Coal Mining

Deep shaft coal mining is different to those mining metalliferous ores. It either operates using the traditional bord-and-pillar method (or room-and-pillar) in which tunnels are mined out along the seam leaving open spaces (bords or rooms) and large section supporting the ceiling (pillars) between them. As the coal is mined out, the pillars are also mined at the end of the workings in an open space called the goaf. Here the ceiling is allowed to fall in as the mining operation retreats. In some mines where the coal seams are wide and horizontal, a longwall technique may be used. This is a long complex system of hydraulic roof supports with a cutting wheel on a track which runs horizontally under the supports with a conveyor belt behind to take the coal to the surface.

Part A: Bord-and Pillar Mining - Great Northern Seam, Lake Macquarie, central New South Wales, Australia. Open the site of the video at:

https://www.youtube.com/watch?v=od87_PVwPP4

Part B: Longwall Mining, United States.
There is no commentary on this video so notice how the system works, especially the excavating cutter and the use of the hydraulic supports above. Open the site of the video at:

https://www.youtube.com/watch?v=WmwEB4DY_jc

The location of this mine is not given, but it is probably in the eastern coal-mining states, Watch the video to the end as it shows how the hydraulic roof supports advance as the coal is removed.

QUESTIONS:

1. What circumstances would determine which type of coal excavation is used?
The orientation and thickness of the coal seam.

2. What type of equipment is used in each method? Compare underground coal mining with that of surface open pit coal mining.
Bord-and-pillar uses low individual coal excavators and shuttles to take the coal to a conveyer belt whereas longwall systems are large and have an integrated system of hydraulic roof supports, cutting wheel and conveyor belt. On the surface, open pit mines can use explosives and large-scaled bucket excavators and haulage trucks to take the coal directly to crushing plants.

3. What are some of the major problems associated with Bord-and-Pillar mining concerning:

    a) construction – bords are mined out in a grid pattern, narrow at first but becoming larger as pillars are also mined.

    b) prevention of collapse - as the bords are mined out, the roof must be supported with mesh and ring bolts using epoxy glue to strengthen the bolts.

    c) working conditions – cramped with machinery and much dust. Also there is the chance of methane (fire damp) explosions so the walls are often coated with white calcium carbonate to reduce any fire.

    d) monitoring of the mine environment and safety – mine inspections usually are common with the trained inspectors testing the air quality for methane and other gases as well as inspecting roof supports and the mine structure and equipment generally. Low voltage is used in electrical equipment to prevent sparks.

4. How is the material exacted and transported from the mine?
Usually the broken coal is taken to the surface by conveyor belt.

CONCLUSIONS:

Summarize the main operations observed in the underground coal mining process. In the summary include the reasons for such mining, the methods used in excavating the mine, any engineering processes designed to hold the mine together, how material is excavated and removed and the main problems associated with such mining.
This could be summarized from a group discussion.

RESEARCH: (Optional)

Use the Internet to find out about the locations of coal mines in the local area, State and Nation. How is this coal transported and marketed and what is its use?

See:

http://www.ga.gov.au/data-pubs/data-and-publications-search/publications/australian-minerals-resource-assessment/coal

http://www.undergroundcoal.com.au/fundamentals/default.aspx

Experiment 19.4: Mining – An Historical Perspective

This is an historical excursion back in time. Most countries, especially Australia and the United States have a rich historical past concerning mining such as the great gold rushes in California and Alaska in the 1840's and the gold rushes in Australia in New South Wales, Victoria, Queensland and later in Kalgoorlie in Western Australia. It is important to understand the life and times of these early mining pioneers, many of whom came to the country from find gold and settled there to redefine the population and culture of their new country.

Part A: Go back in time to the 1851 gold-mining town of Sovereign Hill, Ballarat in the (then) colony of Victoria.

Bendigo and Ballarat were gold-mining frontier towns of the British colony of Victoria in what later became Australia. From a tent city, Ballarat eventually became a prosperous, modern city but in the 1850s things may have been different. Students should watch this video carefully as Sovereign Hill is a faithfully recreated city of the gold-rush era. Students should ignore the tourists and note the buildings, soldiers and townspeople and the old mine and how gold was processed. It is all different today. Open the site of the video at:

https://www.youtube.com/watch?v=xLQUb9tarNA

QUESTIONS:

1. Where is Sovereign Hill located and why did it develop there?
It is located at the city of Ballarat in Victoria, Australia. It has been developed as a tourist attraction on the site of the original goldfields of the 1850s.

2. What was the significance of the 1851 Gold Rush to the development of Australia?
Australia was a penal colony for Britain to send their criminals. This did not stop until 1867 and so a gold rush meant that more free settlers came to the country from many overseas lands, changing Australia from a convict settlement to a country in January 1st 1901.

3. What were some of the differences observed between the lifestyle of the 1850s and today? List all the differences seen or inferred.
Look closely at the dress of some of the characters, the street, the buildings, the soldiers and the mining details.

4. What type of equipment was used in obtaining gold from (a) the surface and (b) underground?
Gold was extracted by shallow digging and panning the streams on the surface but then some miners followed the quartz reefs underground in tunnels held up by timber.

CONCLUSIONS:

Summarize the main features of the colonial town of Sovereign Hill observed in the video and suggest how different conditions were then compared to today. Outline the process of refining the gold using a flow chart.
A good group discussion.

RESEARCH: (Optional)

1. Use the Internet to find out about the Victorian Gold Rush.
2. What contribution did gold rushes make to the countries where they occurred. Discuss.
They brought civilization to many frontier areas and developed the areas into productive parts of a new country.

Part B: Additional Activity: Life on the Victorian Goldfields Game.

This is a simple gaming activity which may give some insights into how people in the early gold-rush days thought and worked. It is a good role-playing activity for any group and different perspectives are offered. It gives players the opportunity to make various decisions and then see what the consequences may be. Good fun and some useful learning.

See also:

https://www.goldfieldsguide.com.au/blog/12/the-victorian-goldfields

http://press-files.anu.edu.au/downloads/press/p198511/pdf/illustrations.pdf

http://www.kidcyber.com.au/gold-rush-in-australia/

https://www.nma.gov.au/defining-moments/resources/gold-rushes

http://www.goldoz.com.au/australian-gold-rush/

http://www.pilotguides.com/articles/colonial-australia-the-gold-rush-and-ned-kelly/

In the United States:

https://www.youtube.com/watch?v=v7PhUMOR99U

https://www.history.com/topics/westward-expansion/gold-rush-of-1849

https://www.larsonjewelers.com/History-of-the-American-Gold-Rush.aspx

https://theconversation.com/how-gold-rushes-helped-make-the-modern-world-91746

19.4 Other Activities

- Visit a local mine. This may be difficult to organise as there are many safety issues. A visit to a mining company is easier to organise.

- Pan for gold (pyrite or galena are good substitutes in sand) using gold pans and pan into large plastic trays or boxes.

- Invite a mine environmentalist to talk about conservation issues in mining.

19.5 Answers to Multichoice Questions

Q1.A Q2.C Q3.C Q4.A Q5.B Q6.A Q7.D Q8.A Q9.C Q10.A

19.6 Some Suggestions for the Review Questions

1. Discuss the factors which could make a mineral one of economic importance. Could a mineral change its status of being economically important or not?
It is a matter of usefulness, supply and expense. Minerals and the metals or the uses obtained from them can change their status if any of these factors change e.g. iron ores are always in demand because of their usefulness and metals such as gold and silver regularly change their value depending upon market forces.

2. Define each of the following terms:

    (a) **cleavage** ability to split smoothly along flat planes of weakness
    (b) **adit** a sloping tunnel into a mine for vehicles
    (c) **metalliferous ore** mineral which is a source of useful metals
    (d) **overburden** unwanted soil or rock removed from the surface in open pit mines to extract the wanted material.

3. Briefly explain, using appropriate mineral/ore examples how each of the following are formed:

    (a) **hydrothermal ore bodies** crystallised in veins or cavities from hot water solutions from a nearby igneous intrusion

(b) **alluvial deposits** valuable material carried downstream in rivers and on beaches
(c) **orthomagmatic deposits** valuable ores concentrated by settling and other processes within large igneous intrusions
(d) **gossans** hard surface cap formed at the top of a vein or other ore body due to weathering of that body at the surface.

4. **Golden specks are seen on the bottom of a shallow, mountain stream. These could be weathered mica, pyrite, chalcopyrite or gold. Briefly outline some observations or tests (assuming that you have the basic prospector's kit) to determine the nature of these specks.**
With a hand lens, the mica will show typical flakes of basal cleavage and the pyrite and chalcopyrite (as sulfides) will react with sulfuric acid (always part of a prospector's kit) to give smelly hydrogen sulfide gas. Gold will be inert.

5. **What are some of the identifiers that a field geologist might look for when prospecting for large deposits of iron minerals (e.g. haematite and magnetite)? What technical assistance could s/he call upon to show the extent and quality of any discovery s/he may make?**
Both minerals often come as huge deposits and typically give the red staining of haematite. Magnetic aerial surveys with a magnetometer will detect the magnetite and density surveys may also help.

6. **Why is the water table - the top of the water level in porous rocks underground - so important to mining? Consider all the possibilities including environmental concerns.**
Water often has a major influence in the weathering of ore bodies but in mining, both surface and sub-surface mines often have a big problem with seepage of groundwater and flooding of the mine workings. Many shaft mines require large scaled pumping even in surface areas which are deserts.

7. **Compare and contrast the possible work hazards associated with open cut mining and underground stope mining.**
Both have hazards associated with the use of mechanized mining, vehicles and explosives as well as work face rock wall collapse. Open cut hazards are more concerned with the first three mentioned whereas underground stopes are very subject to collapse and also air circulation and hazards from gases.

8. **Consider this scenario: A large gold deposit has been located at a depth of 400 metres within jointed metamorphic rock approximately 5 km from a large town. The population of the town is approximately 4000 persons and the surrounding country is mainly used for cattle grazing. The country consists of rolling plains with the occasional rocky outcrop and a small river runs through the town. About one quarter of the country is covered in light forest with an abundance of native wildlife.**

The mining company is discussing whether or not to use open cut, deep shaft or a chemical mining method, such as borehole mining using the cyanide technique. <u>Critically</u> evaluate the potential hazards of the suggested methods prior to giving your decision by:

i. identifying the factors which would be potentially hazardous due to mining
As mentioned above, the hazards concern equipment, collapse and blasting. There will also be the additional problems of dust and toxic wastes going into the pastoral environment.

ii. evaluating which mining technique would be most appropriate for the mine
The type of mine will depend upon the relative ease and extraction of the ore as well as surface environmental concerns such as the relative value of the surface use such as farming and grazing. In this scenario, it may be more useful to mine the ore body by a deep shaft method.

iii. by giving at least two reasons for your choice
Gold in such a geological environment would probably be found within long veins (within quartz) and so it would be better to follow these veins underground than to dig an incredibly large open pit (even at such shallow depth). Also, underground workings would still allow for best use of the surface for farming and grazing.

iv. evaluating the hazards and the environmental concerns
Other than those already mention, there would be problems with surface transport, toxic wastes from any surface processing plant and noise/vibrations from any underground mining. Water table concerns (reduction and pollution) are also an issue.

9. The following values concern amounts of ore mined and the amounts of overburden (unwanted rock) taken out of a surface pit:

Volume of Ore	Volume of Overburden
(Both values are cumulative, i.e. current total.)	
100 cubic metres	30 cubic metres
200 cubic metres	120 cubic metres
300 cubic metres	300 cubic metres
400 cubic metres	540 cubic metres

i. On graph paper, draw a graph showing cumulative overburden related to cumulative ore volumes.
The graph should look something like:

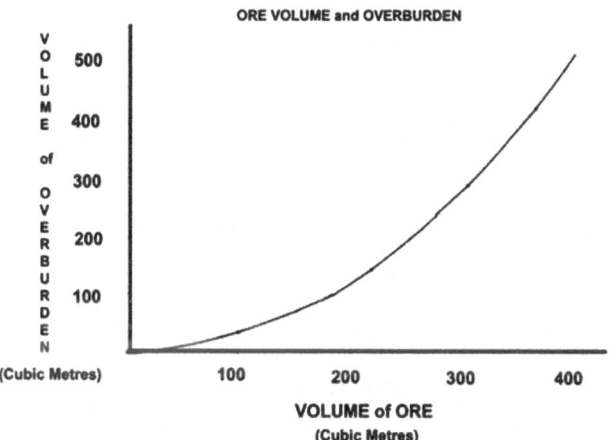

ii. **Explain the trend of the curve and give its significance as far as mining this ore body in the future.**
The amount of overburden needed to be removed to take out the same amount of ore is rapidly increasing. This will mean that mining the ore may become uneconomical.

iii. **How much overburden would have to be taken off to extract 430 cubic metres of ore (assuming that the ore qualities are constant)?**
By extrapolating the graph, it would probably require about 640 cubic metres of overburden to be removed.

10. **What are some of the environmental concerns which should be addressed if an old open pit gold mine was going to be reclaimed, filled and used as a residential area?**
These include: removal of good grazing land, native flora and fauna; dust and overburden on the surface due to the heavy machinery; transportation of ore and overburden through town and rural areas; pollution of the water table due to secondary minerals from the pit and any processing plant nearby (especially from cyanides); and any subsidence or upwelling of toxic waste solutions after the residential area has been built.

19.7 Reading List

Bullock, L. & Hustrulid, W. A. (Edits.). (2001). *Underground Mining Methods: Engineering Fundamentals and International Case Studies*. Littleton, Colo.: Society for Mining, Metallurgy, and Exploration. 718 p. ISBN 0873351932.

Cox, D. P. & Singer, D. A. (Edits.). (2014). *Mineral Deposit Models*. USGS Publication. http://pubs.usgs.gov/bul/b1693/html/bull1nzi.htm.

Dill, H.G. (2010). *The Chessboard Classification Scheme of Mineral Deposits: Mineralogy and Geology from aluminum to zirconium*. Earth-Science Reviews Volume 100, pp.1-420.

Dixon, C. J. (2012). *Atlas of Economic Mineral Deposits.* Springer Science & Business Media. 139 pp.

Guilbert, J. M. and Park, Jr. C. F. (1986). *The Geology of Ore Deposits*, W. H. Freeman. ISBN 0-7167-1456-6

Hartman, H. L. & Mutmansky, J. M. (2002). *Introductory Mining Engineering.* New York. John Wiley & Sons. 570 pages

Mayer, W. (1991). *A Field Guide to Australian Rocks, Minerals and Gemstones.* Sydney: Ure Smith. 335 pp. ISBN: 0 7254 0816 2.

National Environment Agency (NEA): http://www.nea.gov.sg/

Pohl, W. L. (2011) *Economic Geology: Principles and Practice.* New York. John Wiley & Sons. 680 pages.

Chapter 20: Processing the Mined Ore

20.1 Theme

Mining the ore of any useful resource such as metals, sulfur and phosphates is only the first step. Usually the wanted material is physically mixed with gangue (unwanted rock soil and other minerals) and also strongly chemically combined with other elements. Ores and other raw materials of value must be transported to processing plants which may be on site or distant but closer to energy sources, transportation centres or markets.

20.2 Rationale

This chapter deals with the processing of minerals which is a major part of society's industry which turns raw materials into useful products. There are a large number of careers associated with the processing of ores, ranging from the professional metallurgist to process workers within the processing plants. It also involves a great many associated industries and professions which are concerned with the economics of trading in these valuable raw materials. With mining, the processing of ores offers some of the most highly skilled careers as well as the most lucrative.

20.3 Notes on the Practical Work (see PRACTICAL MANUAL 2.)

Experiment 20.1: Separation of a Mixed Ore

This is a simple experiment on separation techniques often given in Junior Sciences classes but it is worth repeating. A good idea is to present the mixture and have students work out the design - perhaps as a flow chart - before attempting the experiment. Ensure that students separate the iron filings from a dry mixture and that the magnet is covered with a waterproof bag. Wet iron filings on a magnet are very difficult to remove. Re-training of how to fold a filter paper may be needed and it is important that the filter paper is not overfilled and dried if it is to be weighed.

QUESTIONS:

1. Why is the mixture used in this activity as ore an unrealistic combination?
It would be unusual to find the compounds together as such a mixture.

2. Why is minimal water to be used in any part of the processes, especially panning?
Because the solution produced will be of low concentration with too much water to be removed by evaporation.

3. During panning, why should the water and solid mixture be stirred?
To ensure that all of the soluble material dissolves.

4. Why was the filter paper wet before that part of the process?
Because dry filter paper does not stick to the glass very well and it will also absorb the initial amount of water and this also helps filtration.

5. What errors could occur at each phase of the separation process? Suggest any improvements to limit errors.
Good student discussion. At every step it is possible that not all of the required material will be separated.

CONCLUSIONS:

1. Give a detailed analysis as a yield % for each of the desired resources.

2. Comment how differences in physical properties have been used in this activity of separation of useful materials.
This will come from student observation but the techniques use magnetic difference (iron removal), density differences (panning), solubility (filtration) and water evaporation.

RESEARCH: (Optional)

Use the Internet to find out how salt is harvested and purified. Why has salt, mined as halite, been an important resource?
Many maritime countries obtain salt from very large evaporation ponds using seawater but some countries in Europe have large, underground salt deposits which have been mined for thousands of years. See:

https://eusalt.com/salt-production

http://www.madehow.com/Volume-2/Salt.html

Experiment 20.2: Froth Flotation

One of the main separation methods used in the separation of metal sulfide minerals such as sphalerite (zinc) and galena (lead) is froth flotation. These minerals are often mixed together and are difficult to separate by other means. IN this simple variation, iron filings and sand are used for convenience as the usual froth flotation makes use of some complex and toxic chemicals. Kerosene (paraffin in the US) is relatively safe but still inflammable and its fumes are a little toxic. It should not be ingested.

QUESTIONS:

1. What is the purpose of the kerosene?
To act as an organic base to separate from the water.

2. What is the purpose of the detergent?
To provide the froth to rise up on the kerosene.

3. How could the iron filings be collected from the test-tube?
By scooping off the bubbles and drying the mixture. If the kerosene and soap still persist, the mixture could be washed in alcohol which should remove the organic component and when this evaporates the iron should be purer.

4. Why is such a technique important to the mining of metals?
It is a quick way of separating ores efficiently and in large quantities.

CONCLUSIONS:

Summarize the method and observations made during this experiment.
Student summaries and group discussion about effectiveness.

Also see an animation explaining how the technique works at:

https://www.youtube.com/watch?v=pFbiatPD4ZM

RESEARCH: (Optional)
Use the Internet to find out how froth flotation is used in other industries.

https://www.sciencedirect.com/topics/chemical-engineering/froth-flotation

https://www.researchgate.net/publication/258712096_The_use_of_froth_flotation_in_environment_protection

Experiment 20.3: Chemical Extraction of Copper from Copper Ore

This is a good experiment to show both the need for physical separation and several complex chemical reactions are needed to obtain a material in a relatively pure state.

CARE: Sulfuric acid is corrosive and copper solutions are poisonous. Both should not be handled and care should be taken with the eyes. It is strongly advised that students wear appropriate safety equipment such as aprons, rubber gloves and safety glasses. In addition, all materials used in the experiment should be emptied into a class waste bucket for later removal. Copper sulfate solution can be flushed down the sink with plenty of water but all other materials should be dried and then wrapped in plastic and placed into the rubbish bin.

The copper ore, a mix usually of malachite (green) and azurite (blue) is soft but it may be impregnated within a hard rock. In lieu of an industrial rock crusher, small pieces of ore can be crushed in cloth bags on the floor using relatively gentle hits with a geology pick.

QUESTIONS:

1. What is property of the azurite/malachite ore which enables it to be separated from the gangue mixed with it?
They are both carbonates and so will react with an acid to form a solution.

2. What is the purpose of the iron nails?
Iron is more reactive chemically so it will replace the copper from the copper sulfate solution. Not as efficient as electroplating but it works well.

3. What is the chemical property of iron which enables this to happen? What other metals could be used instead of iron? (Hint: see the reactivity series of metals).
Iron is more reactive than copper. See:

https://www.thoughtco.com/activity-series-of-metals-603960

4. Why is such a technique important to refining of copper?
It enables a good yield.

CONCLUSIONS:

1. Summarize the method and observations made during this experiment.
2. What is the percentage yield produced by this experiment?
3. What are some of the probable errors which have occurred? How could they be reduced?

These questions will come from the experiment but the yield will not be too high because it relies upon full reaction with the acid and some copper carbonate will still be inside the crushed ore and unless excess acid is used it might not also fully react. The copper coating on the iron also depends upon the surface area of the iron which some becomes plated. Perhaps some small sheets of iron may be available. Further crushing of the ore and excess acid would assist the reaction and more iron nails or a bigger surface will improve yields.

RESEARCH: (Optional)

Use the Internet to find out how copper ore is processed and the copper metal refined.

See:

https://superfund.arizona.edu/learning-modules/tribal-modules/copper/processing

https://www.copper.org/education/copper-production/

http://www.essentialchemicalindustry.org/metals/copper.html

20.4 Other Activities

- Separate a magnetic ore (e.g. magnetite or ilmenite or monazite) from non-magnetic ore (e.g. crushed grey haematite) or separate the magnetic component (ilmenite/monazite) from heavy mineral sands. Use a bar magnet wrapped in a plastic bag so that the particles adhering to it can be removed.

- Play the Placer Mining Game at:
  http://www.see.leeds.ac.uk/misc/miner/

- Interactive mining site at
  http://www.oresomeresources.com/media/flash/interactives/minerals_downunder/

- Silver plating copper (demonstration). Place the copper plated nails from the experiment or obtain some copper sheets. Place the copper article into a test-tube of silver nitrate solution. CARE is needed as the silver nitrate will react with salt in the skin to produce silver chloride which instantly reacts with light to give black free silver. These dense black stains last a long time! After a while the silver plating can be shaken off and students can see the shiny silver. Another good example of a purification by replacement from a solution.

20.5 Answers to Multichoice Questions

Q1.A Q2.C Q3.D Q4.B Q5.C Q6.D Q7.A Q8.B Q9.D Q10.A

20.6 Some Suggestions for the Review Questions

1. Discuss the factors which could make a mineral one of economic importance. Could a mineral change its status of being economically important? How

This question has been asked before but it is important to relate it to economic mineral processing. Over times, some processing systems become expensive e.g. aluminium and copper processing requires a very large amount of energy and so smelters are often located near the source of relatively cheap electricity – near hydroelectricity systems or near coal-fired or nuclear power plants. Usually the raw ores must be transported long distances. Sudden increases in energy costs or decreases in supply, especially in coal or oil-fired energy regions, may reduce production in favour of other materials.

2. Define each of the following terms:

   (a) **ore** mineral of economic importance
   (b) **gangue** unwanted minerals mixed with ores
   (c) **slag** waste product of silicates removed from smelting
   (d) **smelting** extraction of metals using heating, melting and usually chemical change.

3. A sample in a 2 kg bag, brought in from exploration in the field, consists of rocks containing galena, magnetite and sphalerite.

Outline a laboratory sequence that may be used to separate the ores from the sample.

1. Whilst in the solid mix, magnetite can be extracted with a bar magnet wrapped in plastic which is stirred around in the mix.

2. Next, the galena (SG = 7.5) and the sphalerite (SG = 4.0) can be separated by density differences. This can be done using a small-scale froth flotation cell consisting of a stirring propeller within a small container and a tube to pass air. Some kerosene and detergent is added with the finely powered minerals. The mix is then frothed up and the lighter sphalerite should float to the surface on bubbles and be scooped off. A similar activity is found at:

   http://www.oresomeresources.com/media/flash/interactives/oresome_froth/assets/pdf/teacher_experiment.pdf

4. How is bacteria used in the extraction of metals from soil and gangue? Research the Internet to find out how these bacteria is grown.
Some bacteria such as *Theobacillus ferrooxidans* lives off the chemical energy trapped in metal sulfides such as copper sulfide found in ores or easily made from oxides. The bacterium then removes the copper and sulfur as waste products. Some other bacteria also extract gold from toxic gold solutions – see:

https://blog.nationalgeographic.org/2013/02/06/scientists-discover-how-bacteria-changes-ions-into-gold/

5. Housing estates are sometimes built on old gold fields. What would be some of the hazards which could be associated with living in such as estate?
In the past, some estates in major cities (e.g. Brisbane, Australia) have been built on top of old gold workings. Many of these in the past when reclamation laws were few, still contained old shafts, near-surface tunnels and toxic waste dumps from the gold processing plant. Older processes often used cyanide and mercury compounds which were left behind and superficially covered. Changes in the water table or construction digging often exposed these toxic chemicals or caused local collapse of the surface.

6. How is recycling used in the metals processing industries? Are metals recycled in your society? Why should metals be recycled?
Scrap metal is a major industry and even gold from computer terminals and other electronic equipment is recycled. Most of the major metal refineries use a good percentage of scrap metal which is placed into the furnace alongside the raw metals produced by initial smelting. Making steel from recycled cans for example uses 75% less energy than when producing steel from raw materials.

7. Research the internet to detail the metals obtained from the following ores:

(a) **haematite** iron
(b) **bauxite** aluminium
(c) **rutile** titanium
(d) **sphalerite** zinc
(e) **galena** lead
(f) **chalcocite** copper
(g) **magnetite** iron
(h) **monazite** rare earths & thorium

8. Processing plants are often located in specific regions which may be a considerable distance from the mines which produce the ores which are processed and so the ores must be transported over these long distances. Explain this apparent inefficiency in regard to:

    (a) aluminium production
    This requires a very large amount of electricity so the smelters are located near power stations
    (b) iron production
    This requires a large amount of coke (carbon) and so smelters are located near coal fields and also near sources of limestone.

    What is the main factor for the location of such plants in these examples?
    Usually the energy factor or the bulk material needed to be added to the concentrated (and less volume) ore.

9. Mercury metal was once the main processing agent in extracting gold from crushed gold-bearing quartz. Why was it used, and what were the hazards associated with this process? What are the environmental hazards associated with the smelting of iron, copper and aluminium?
Mercury would dissolve gold and form an amalgam so it could be easily separated from the lighter soil. The gold is then extracted from the amalgam by distillation which vaporized the mercury and left the gold. Unless this distillation system was completely sealed, some very toxic mercury vapour would escape to the environment.

10. Do an audit of your home or educational institution for the ways that processed metals are used. What could be some substitutes which could be used for metals if they became scarce?
This is a very useful exercise as our modern society has become dependent upon the use of metals in many subtle ways. See:

    http://www.diydoctor.org.uk/projects/metal_used_at_home.htm

During times of war, when metals were required for war material, some of the older materials such as wood and stone would have to be used for construction materials and earthenware for utensils.

20.7 Reading List

Davenport, W. G., King, M., Schlesinger, M. and Biswas, A.K. (2002). *Extractive Metallurgy of Copper, 4th Edition,* (Pergamon Press: Oxford, England), 91-102.

Fuerstenau, M.C. & Han, K.N. (2003). Principles of Mineral Processing. Dearborn, MI: SME. 573 pp. ISBN-10: 0873351673.

Pohl, Walter L. (2011) *Economic Geology*: Principles and Practice. New York. John Wiley & Sons. 680 pp. ISBN: 978-1-4443-3663-4

Wills, B. A., Napier-Munn, T.J. (Edit.) (2006). *Wills' Mineral Processing Technology, (Seventh Edition): An Introduction to the Practical Aspects of Ore Treatment and Mineral Recovery,* 7th Edition. Oxford: Butterworth-Heinemann. 444 pp. ISBN 9780750644501.

Additional References

https://www.youtube.com/watch?v=6kFONdchY0U Another froth floatation video

http://www.oresomeresources.com/resources_view/resource/publication_the_science_of_mining/section/resources/parent/rocks_minerals/category/smelting_refining smelting

http://www.oresomeresources.com/resources_view/resource/movie_copper_refining_plate_removal/section/resources/parent/rocks_minerals/category/smelting_refining copper refining

http://www.oresomeresources.com/resources_view/resource/experiment_froth_flotation/section/resources/parent/rocks_minerals/category/smelting_refining
Froth flotation

https://www.youtube.com/watch?v=9N6uXQ8KRYc good animation of iron production
https://www.youtube.com/watch?v=og-Pzzf2zdM good video of history of iron production

https://www.youtube.com/watch?v=fa6KEwWY9HU Aluminium production

https://www.youtube.com/watch?v=pD3Nax8F3Hk More on Aluminium

http://www.digintomining.com/high-school-resources
some mining teaching resources
https://www.911metallurgist.com/blog/lead-zinc-flotation-separation-method
Good Lead-Zinc fact sheet with video

Chapter 21: Monitoring and Management

21.1 Theme

Regular monitoring of processes and the environment has always been a part of the industrial and commercial world as part of the management system. In more modern times, pollution of the spheres of the Earth has led to monitoring of the natural environment so that polluting processes and their effects can be controlled and where possible, minimised or removed.

21.2 Rationale

The realisation world-wide that all of the Earth's spheres need to be monitored and better managed has come slowly and at a price. There is still some resistance to the changes that need to come about at both the national and international level. Resource companies, governments and sometimes populations wish to maintain the past of having available resources, relatively cheap energy and a good standard of urban lifestyle.

21.3 Notes on the Practical Work (see PRACTICAL MANUAL 2.)

Experiment 21.1: Monitoring of Local Climate

This is a general, long-term experiment using standard weather instruments and revises previous activities about such instruments. For those institutions which do not have a designated weather station, one can be simply established from this activity. With the purchasing of a simple aneroid barometer, students can use thermometers and make a simple weather station at home. Electronic weather stations are freely available at a reasonable cost. After recording the parameters for at least a week, students may see a trend and be able to make predictions. Their readings can be compared nightly on TV weather reports and on the many weather Apps. available on mobile phones and computers. See:

https://www.instrumentchoice.com.au/instrument-choice/weather-stations/home-weather-stations/0-100-weather-stations

https://the-weather-station.com/

http://www.e-missions.net/wvstorm/?cat=2&sid=1&pid=31&page=Understanding%20Weather%20Measurements%20To%20Predict%20Weather

QUESTIONS:

1. Has the weather changed over the month? How (give details).
2. Was the weather in the next week as predicted? Comment.
3. Has the average for that month temperature changed locally over time? (give details).

The answers will depend on the local weather conditions over this time.

4. List all of the problems or errors which could occur with such short-term weather measurement.

The answers to these questions will come from student measurements and evaluations. Accuracy of short-term measurements sometimes are in doubt for all of the reasons of any short-term sampling as most weather predictions are based upon long-term average trends and computer algorithms.

CONCLUSIONS:

Comment on the usefulness of monitoring local climate and any changes observed over the (a) short term and (b) long term.

See the notes above in reference to short-term measurement. Long-term measurements and trends for the basis of professional weather predictions and usually from data received from many places over a wider area.

RESEARCH: (Optional)

1. How is weather data gathered on a wider scale?
2. How are weather predictions made?

See:

http://www.bom.gov.au/climate/data-services/education.shtml

https://research.csiro.au/dfp/wp-content/uploads/sites/148/2018/05/BOMs-Climate-forecast-Verification-Wang.pdf

Experiment 21.2: Monitoring of Carbon Dioxide Levels

This experiment revises the simple concept of carbon dioxide in the air and from aerobic respiration and then goes on to examine world-wide monitoring of carbon dioxide levels.

Part A: Carbon Dioxide in the Room

This simply shows that carbon dioxide exists in air and also comes from aerobic respiration in higher amounts. Limewater is again used as the indicator and a control test-tube is stopper to show the effect of air and the breath on the limewater.

Part B: Carbon Dioxide Levels in the Atmosphere Worldwide

This part of the experiment shows how carbon dioxide is monitored world-wide and enables students to obtain data and then use it to see if there has been a world-wide trend in carbon dioxide levels. The graph should show the levels have risen rapidly from the 1950s to over 400 parts per million today.

Open the following websites which gives daily CO_2 readings from the observatory at Mauna Loa, Hawaii:

https://www.co2.earth/daily-co2

Bookmark this site so that regular readings could be taken in the future.

https://data.giss.nasa.gov/modelforce/ghgases/Fig1A.ext.txt

QUESTIONS:

1. What is the current level of carbon dioxide gas at Mauna Loa in parts per million?
Readings today (Dec. 28, 2019) are: 411.60 ppm (Scripps) and 408.26 ppm (NOAA)

How does this agree with the extrapolation? Explain any differences (+/- 10%).
This will depend upon the accuracy of student graphs.

2. Why are the values taken from measurements at Mauna Loa in Hawaii?
Hawaii is a long way from major industrial and commercial sources of carbon dioxide and it is high enough to get an uncontaminated value. However, there may be some problems at times with the local volcanic smog or vog.

3. Was there any change in the recorded CO_2 levels over the month? Explain, giving details of any possible errors in sampling.
This will depend upon the data and the season. See:

https://www.esrl.noaa.gov/gmd/obop/mlo/programs/esrl/co2/co2.html

https://journals.ametsoc.org/doi/pdf/10.1175/EI224.1

CONCLUSIONS:

Make a general comment on the change in CO_2 levels over the last sixty years and give some explanation for any changes.
Sudden increase due to fossil fuels and other man-made process emissions.

RESEARCH: (Optional) How and where are these measurements made and what errors could be involved in measuring carbon dioxide levels?
See the websites given above and:

https://pdfs.semanticscholar.org/af5c/95f628d27806ae16f87b97b9949582ce5f1e.pdf

https://hal.archives-ouvertes.fr/hal-00295825/document

Experiment 21.3: Monitoring of Air Quality

This is an extension of the previous experiment and introduces students to the total monitoring of the air of the atmosphere and he use of the Air Quality Index (AQI). This experiment uses another website which should be bookmarked to give further values in the future for personal use. Open the following website for Australian air quality values:

http://aqicn.org/here/

In the United States, real-time AQI for different locations can be found at:

https://aqicn.org/map/usa/

AQI values for other locations can be found by typing the nearest city into the search box or the other useful site is found at:

https://waqi.info/

In **Part A: Local and national air quality**, values and changes will depend upon the students' locality, the climate and any local industry.

In **Part B: International values of AQI**, some interesting comparisons can be made with cities and countries which must live with high values of poor air quality because of their pollution levels. Generally speaking, countries in the Southern Hemisphere such as Australia enjoy regularly low values of AQI but in the Northern Hemisphere, some cities have high values for most of the time with only some rare exceptions of clean air.

QUESTIONS:

1. What locations had (a) high values and (b) low values of AQI?
As suggested above, places like cities in China and India have very high values of AQI whereas places with less pollution such as Australia and New Zealand enjoy low values AQI. The last website is excellent as it gives a map which shows extreme and low values by colour on a world map.

2. What were the main individual AQI pollutants in places with high values? Give probable reasons for these high values.
This will vary from place to place but there can be high values of particulate matter, sulfur dioxide, ozone, carbon monoxide and nitrogen dioxide.

3. What factors would vary such high values from time to time?
These will change with changes in industrial pollution and climate. During days when transportation is limited in the city values will go down. They also go down after rainfall and when fresh winds blow into the region. There is also seasonal change with temperature.

CONCLUSIONS:

Make any comment generally about the international AQI values and reasons for major differences observed across world locations such as climate or other factors.

These will come from student discussions but will be similar to the views given above.

RESEARCH: (Optional):

1. How and where are AQI values calculated? What are their usefulness and limitations?

https://app.cpcbccr.com/ccr_docs/How_AQI_Calculated.pdf

https://stimulatedemissions.wordpress.com/2013/04/10/how-is-the-air-quality-index-aqi-calculated/

2. Name some places which have extremely high and dangerous values of AQI?

Dangerous places occur from time to time in some of the Chinese, Indian and other Asian cities where occupants are advised not to go out. In general, when visiting these countries or others suspected of having high AQIs, asthma sufferers, older people and children should take precautions.

See:

https://www.epa.gov/sites/production/files/2014-05/documents/zell-aqi.pdf

https://airnow.gov/index.cfm?action=airnow.calculator
(a good calculator with warnings)

Experiment 21.4: Radon Gas Levels

Radon gas is an inert radioactive gas related to the other noble elements, helium, neon, krypton and xenon. It naturally comes off radioactive isotopes in crystalline rocks such as granite. Normally it would not be a problem in nature, but in cities with basements and other low-lying places with poor circulation, radon gas can accumulate as it is denser than air. Its radioactivity can then cause cancers in people who live or work in these areas. This is why monitoring of radon can be important. The first website is excellent as it is interactive and will give the actual values in Becquerels per cubic metre even when zoomed into to individual suburbs of cities. In general, colours are also used to give concentrations with blue being low and yellows high.

http://arpansa.maps.arcgis.com/apps/Embed/index.html?webmap=c7501ea15f45467da37059b21ec8e66e&extent=105.5019,-42.4303,167.2451,-11.9624&home=true&zoom=true&scale=true&search=true&searchextent=true&legend=true&basemap_gallery=true&theme=light

For users of the black-and-white editions, the EPA map for radon levels in the United States can be found at:

https://en.wikipedia.org/wiki/File:US_homes_over_recommended_radon_levels.gif
There is also an interactive map at:

https://www.epa.gov/radon/find-information-about-local-radon-zones-and-state-contact-information#radonmap

Radon and other maps concerning radiation for Europe can be found at:

https://remon.jrc.ec.europa.eu/About/Atlas-of-Natural-Radiation

World radon levels can be found at:

http://www.sbpr.org.br/pdf_sbpr_2014/44_JAN_IISBPR_2014.pdf

QUESTIONS:

1. How does the local value of radon emissions compare with that of the average for Australian urban centres?
This will depend on the locality.

2. Comparing the maps for the United States and for Australia, what can be said about the average radon emissions for each country?
Australia generally has a much lower rating as it has a warmer climate and little need for basements.

3. Where are the major places for greatest radon emissions in Australia?
Higher readings for radon are in places which have granites and similar rocks beneath e.g. the Snowy Mountains.

CONCLUSIONS:

Australia generally has lower values of radon emission so it is not considered of great importance except in individual places e.g. some tunnels in the Snowy Mountain Scheme.

RESEARCH: (Optional)

1. Research the life and work of Henri Becquerel.

https://www.atomicheritage.org/profile/henri-becquerel

https://www.biography.com/people/henri-becquerel-40055

2. What workplace health and safety issues are there in mining and prospecting for uranium?

Miners and prospectors are usually very close to radioactive sources if looking for that resource. Radioactivity problems are cumulative so time is also a factor. In mines such as the uranium/copper mine at Olympic Dam, South Australia, radon and other radioactivity levels must be monitored and appropriate ventilation of the mine is done.

Experiment 21.5: Water Quality Monitoring

This experiment was designed in lieu of a full professional water-monitoring kit that may be available. There are a great range of water-testing kits suitable for class use. These include:

http://www.lamotte.com/en/education/water-monitoring

https://www.carolina.com/environmental-science-water-quality/lamotte-green-water-quality-monitoring-kit/652567.pr

https://www.selectscientific.com.au/water-quality-test-kit.html?gclid=EAIaIQobChMIqLetmuaU4AIVUI6PCh2yswHzEAMYASAAEgJO5PD_BwE

https://www.amazon.com/Lamotte-GREEN-Program-Water-Monitoring/dp/B00BWXI1L4

However, this experiment uses convenient laboratory equipment to carry out some water tests as examples of what can be done in environmental monitoring. The tests can be done on water samples brought in by students (CARE: remind them in advance about dangers of going near waterways – always go in a group). Some water-testing strips such as pH and chloride can be obtained cheaply from swimming pool suppliers.

PART A: pH

This is simple and uses Universal Indicator to measure acidity levels

PART B: Water hardness

Hardness (to lather soap) is often a common problem in some area. The test can be done on natural water to see if soap will lather and also in pre-made hard solutions of temporary (using Ca(HCO$_3$)$_2$ and Mg(HCO$_3$)$_2$) and permanent (SO$_4^{2-}$ and Cl$^-$ ions of Ca and/or Mg). Boil the temporary hard water and lather and use a water conditioner (e.g. washing soda) to remove the permanent hardness. Explain the use of these in domestic washing and also the use of modern detergents instead of soap.

PART C: Test for Chlorides

Using silver nitrate will give a cloudy precipitate of silver chloride which is photo-sensitive and turns purple in sunlight due to the presence of a colloid with black free silver. Take care as the silver nitrate will stain many things a permanent black – especially skin!

PART D: Test for Sulfates

Another simple water test which can be done on local tap water, stream samples or a pre-made sodium sulfate solution. A dense white precipitate of barium sulfate forms.

PART E: Oxygen content:

This requires the use of indigo carmine solution which may be obtained from the Chemistry or Biology Faculty. It is 5,5'-indigodisulfonic acid sodium salt often called Acid Blue 74 and is sold as a blue food colouring for animals so look for it in the pet shop or local pharmacy. It can be purchased online at:

https://www.melbournefooddepot.com/buy/indigo-carmine-powder-20g/F01221

http://astralscientific.com.au/indigo-carmine-10g-indicator.html
https://www.wardsci.com/store/product/8877586/indigo-carmine

The safety information of this dye is given at:

https://efsa.onlinelibrary.wiley.com/doi/pdf/10.2903/j.efsa.2015.4108

Test a sample of ordinary pond water or even tap water and match the colours against the following values (in parts per million - ppm):

> Yellow 0.000
> Orange 0.005
> Orange-Pink 0.010
> Pink 0.015
> Pink-Red 0.025
> Red-Purple 0.050
> Purple 0.100

Use the same sample and boil the water to remove the air and place a stopper on the test-tube to prevent any air dissolving in the water. Retest the cold, de-oxygenated water.

Watch the video at: https://www.youtube.com/watch?v=_e8ENtdBmlc

The video shows the neat traffic lights colour change experiment and the recipe can be found at:

http://www.chem.ed.ac.uk/sites/default/files/outreach/experiments/indigo-tech.pdf

Part F: Nitrates – Demonstration of the Brown Ring Test

This is another standard analytical chemistry test. It should be given as a demonstration on water to which a little ammonium nitrate has been added then try a local water sample. It would be ideal to have some water sampled from runoff from an agricultural field to see if there is any water pollution. Take care using the concentrated sulfuric acid so use gloves, apron and glasses. It can be explained that this is a laboratory test and that field equipment is a lot safer.

QUESTIONS:

1. Why was the sample bottle originally washed, rinsed and dried before collecting water?
So that any water samples are not contaminated.

2. What is the usual pH level preferred by most organisms?
Usually between pH 6 and 8 but some extremophiles can go to low pH values.

See:

http://www.ei.lehigh.edu/envirosci/enviroissue/amd/links/wildlife3.html

3. Where do the calcium, magnesium, hydrogen carbonate, sulfate and chloride ions in water come from?
They have been leached out as mineral salts from minerals in soil and rock.

4. What are some of the main errors in the methods used here for water sampling?
The main error with any chemical testing of small concentrations is the sensitivity of the test, the freshness of the solutions and the concentration of the ions being tested in the water sample. Remember that some of the water tested in this experiment may not contain the ions being tested.

CONCLUSIONS:

Make a list showing the name of the test, what property/substance it tests for and the result if positive.
From group discussion assuming that a positive test can be seen or demonstrated.

Discuss the need and usefulness of water testing for (a) domestic water supplies, (b) the local ecology, (c) industry and (d) mining (some research may be needed).
In general, many water ecologies have very limited ranges of temperature, pH, oxygen and mineral salt levels so these need to be monitored if the environment is to be protected from chemical wastes, eutrophication due to algal blooms and general water pollution from domestic, agricultural and industrial runoff.

RESEARCH: (Optional)

Research how water quality is monitored in the oceans See NOAA sites)

See:

http://www.aoml.noaa.gov/outreach/floridaseagrant/pdf_files/TropicalConnections_WaterQualityIsMonitoredToAssessEnvironmentalConditions_KelbleHeilGlibert.pdf

https://www.nnvl.noaa.gov/StoryMaps/DITC/WaterQuality/WaterQualityTeachersGuide2016.pdf

Experiment 21.6: Total Dissolved Solids - Demonstration

Total Dissolved Solids (TDS) is the sum of the cations (positively charged) and anions (negatively charged) in water and provides a qualitative measure of the amount of dissolved substances in the water. TDS monitoring is important in the processing of drinking water and in many industrial processes where dissolved materials could cause harm to the equipment if they solidify.

PART A: Gravimetric Analysis

The simplest way to measure TDS is to weigh a large, empty container, fill it with a known volume of solution having dissolved solids, boil it dry and then reweigh the container to find the mass of the solids and therefore the concentration in the original volume. This will take some time but shows the traditional method. The time factor is one of the disadvantages of this experiment and in other cases, if the TDS contains chemicals which are heat sensitive, they may decompose after the water has boiled off.

The use of the spectroscopes is an added activity to show that the solids obtained by this method can then be analyzed for composition by other means. If a strong salt solution is used (recommended) then the remaining solid will show the bright yellow flame for sodium in a burner and the two bright orange lines for sodium with the spectroscope. A sample of the solution will also show the presence of chloride ions if silver nitrate solution is used. This is a good example of analysis provided that the students do not know what was in the original solution.

PART B: Conductivity Estimation of TDS

This is a good experiment to show that in monitoring situations, a quick result with good accuracy can be obtained by portable, digital means such as a TDS meter to be used in the field.

Make sure that the solution used in this part is the same as that boiled dry in PART A. Here a common multimeter was used and a simple tank which can be measured in all three dimensions to get a standard unit which can then be converted to real values of concentration. A trial of this experiment gave an accuracy of less than 10%.

QUESTIONS:

1. What was the instrument error of (a) the multimeter, (b) the measured dimensions of the tray and (c) the area of one end?

The error of the digital multimeter will be that of the smallest unit shown on the scale, the dimensions of the measured dimensions will be +/1 one half of the smallest unit (i.e. 0.5 mm) but if the units are multiplied then one should add the errors.

2. What other errors could occur in this estimation (itself a very rough method)?

The general sensitivity of the multimeter and student measurement errors.

CONCLUSIONS:

1. What was the value for the TDS determined by gravimetric analysis (accurate)?
2. What was the estimated value of the TDS using conductivity?
3. Comment on the differences between both values and any errors involved.

These answers will come from how the experiments are performed. If done with care, the estimation will come close to the value for boiling off the water.

4. Comment of the pros and cons of using both methods

The first method is bulky and takes a considerable time and energy and would be done later in a laboratory. The electronic method using a proper TDS water testing device is only an estimate but is portable, sturdy and able to give instant results in the field.

RESEARCH: (Optional)

Use the internet to find out how the TDS (dry) could be analysed to find out its exact composition.
Hint: look for sites dealing with flame spectrometers and other chemical analytical methods.

Experiment 21.7: Turbidity Management

Secchi disks have been mentioned before. This experiment uses small, home-made disks which can be easily used in simple samples in the laboratory.

PART A: Sedimentation and settling

This is a very simple but informative activity which students seem to enjoy. The settling of larger particles happens very quickly if normal gravel, sand and silt are used. It may take 24 hours or more to the contents of the bottle to settle into distinct layers and give a clear solution above. It will be noted that some of the sand will get between the gravel to give a good representation of future conglomerate. Watch the author's video at:

https://www.youtube.com/watch?v=yaYO4lc_G3M

The second part of the experiment uses mini-Secchi disks made from a metal washer with the Secchi pattern drawn (with <u>waterproof</u> black ink) on paper and glued to the washer. The paper will deteriorate with several dunkings so use a water-proof white cardboard or water proof both sides with some clear adhesive contact so that they can be used again.

QUESTIONS:

1. How did the sediment settle immediately after the shaken bottle was put down?
The gravel settles almost immediately and the sand very quickly and the coarser silt soon after. Finer silt takes some time.

2. How long did it take for the coarser sediments to settle compared to the fine, suspended sediments?
As above.

3. What factors will determine the rate of settling of sediments in water?
The weight and size of the particles (size affects the sinking rate), the viscosity of the water (minor but due to other sediments) and the initial force of disturbing the sediment by shaking.

4. Did the Secchi dish prove effective in measuring turbidity?
5. What was the estimated time (by graph extrapolation) for the suspended solid to clear?
These two questions relate to the student observations but under good conditions any extrapolation for expected clarity should work out.

6. Was there this clarity in the bottle set aside over night? Comment.
This will depend upon the fineness of the silt. It might take longer.

CONCLUSIONS:

Using the bottle as a model, comment on the use of settling tanks to reduce turbidity in waterways and industrial sites.

Comment on the effectiveness of chemicals such as aluminium sulfate (Alum) in reducing turbidity.

From group discussion it should be seen that settling in ponds is a good idea but it takes time and relies upon still water conditions. A flocculant such as the aluminium sulfate is useful provided that it is not considered a contaminant,

RESEARCH: (Optional)

1. Use the internet to find out how sedimentation ponds are constructed.
2. What other methods can be used to remove sediment from water on a large scale?

See:

https://www.catchmentsandcreeks.com.au/docs/SB-1.pdf

https://www.austieca.com.au/documents/item/698

http://www.eng.auburn.edu/files/centers/hrc/ir1301designandconstruction.pdf

Turgid water can also be cleared by filtration systems.

21.4 Other Activities

- Repeat experiment 21.6 b for different known concentrations, calculate the TDS value and hence calibrate a simple multimeter as a TDS meter for field use.

- Take an excursion to a local waterway to collect samples and take field measurements. Do a Risk Assessment first.

- Invite a member of the local government or private environmental protection office or national park service to talk on water and other environmental monitoring.

- Use a light meter to monitor light conditions in various parts of the local environment and relate this to plant growth if possible. See:

    https://www.wikihow.com/Measure-Light-Intensity

21.5 Answers to Multichoice Questions

Q1.A Q2.C Q3.D Q4.A Q5.C Q6.A Q7.B Q8.C Q9.D Q10.B

21.6 Some Suggestions for the Review Questions

1. **Research the links given at the start of this Chapter and the Internet in general to list the main environmental agencies of:**

    (a) this country; and
    (b) the world.

    See:

    http://www.environment.gov.au/about-us/international
    (Australian Department of Environment and Energy)

    https://www.epa.gov/
    (US Environmental Protection Agency) also see NOAA and NASA sites.

https://www.eea.europa.eu/
(European Environment Agency)

Some of the world's environmental groups of significance can be found at:

http://www.ecology.com/government-environmental-agencies/

https://www.unenvironment.org/civil-society-engagement/accreditation/list-accredited-organizations

2. **Think of some reasons why individuals, private companies and governments do not often pay much attention to environmental research. Discuss this as a group exercise.**
   This depends upon the outcome of the discussion, but one might look at ignorance, indifference, profits before environment, lack of funding. Also research where environmentalists are employed in research in the private sector.

3. **Distinguish between primary and secondary air pollution and give an example of each.**
   Primary air pollution is that emitted directly from a source e.g. from burning fossil fuels or from a volcanic eruption. Secondary air pollution forms when primary pollutants react in the atmosphere including smog, ground-level ozone and acid rain. See:

   https://soe.environment.gov.au/theme/ambient-air-quality/topic/2016/pollution-types

   http://www.differencebetween.net/science/difference-between-primary-pollutants-and-secondary-pollutants/

4. **List the main anthropogenic sources of pollution of the:**

   (a) Atmosphere – burning fossil fuels
   (b) Hydrosphere – runoff from the land including sewerage and agricultural runoff from rivers.

5. **Discuss the amount of waste which is put into the local environment and suggest some ways that it may be used as a raw material.**
   This depends upon the students' knowledge of waste in the local environment. Recycling of metal, glass, paper, tyres etc. and use of waste dumps and furnaces as energy sources would be major discussion points.

6. **Why is particulate pollution a problem in the atmosphere and the hydrosphere? Give some examples of problems which can occur with particulate pollution.**

   Mostly a health problem with respiratory irritation as well as environmental concerns when the particles interact with the natural systems e.g. volcanic ash destroying forests and crops as well as industrial processes which rely upon clean air. See:

   https://www.epa.gov/pm-pollution/health-and-environmental-effects-particulate-matter-pm

7. **Revise the need for accuracy in measurement when it comes to instruments which are used to monitor the environment. What is calibration and why is it needed?**

   Accuracy is important in giving validity to any scientific hypothesis or judgement e.g. no one will take an environmentalist serious if they simply say that an ecosystem is polluted. Hard, facts from good, reliable and accurate measurement is required to say how the ecosystem is polluted, in what concentrations, where and by what pollutant. Calibration is important to prove that measurements are valid and really measure what they claim to measure. Too many environmental activists often do not backup their statements with hard facts and scientific arguments. The way forward in environmental activism (sometimes necessary) is to publicly present the case based on sound, reliable evidence.

8. **What is radon? How would it get into an urbanised area and how can it be measured? What long-term harm would high radon counts have on a community? (Some extra Internet research).**

   Radon is a heavy, inert (chemically inactive), radioactive gas which pools in underground areas such as mines, basements and underground carparks, especially in countries with sub-surface granites and similar rocks. In those countries with harsh climates, especially cold winters, people are inclined to use basements and shelters which may have high radon levels. Radon is known to cause cancers, especially lung cancers from breathing in the gas.

9. **Some animals are homeothermic and some animals are poikilothermic. What does this mean and what are the implications for these animals with increased global warming?**

   Poikilothermic (varied temperature) animals do not maintain a constant body temperature but have a body temperature which varies with that of the outside e.g. fish, amphibians and reptiles. Homeothermic (same temperature) animals are able to maintain a constant body temperature because of energy from respiration and body insulation such as blubber, hair, fur and feathers e.g. birds and mammals. With increased global temperature, poikilothermic animals, those animals which find the environmental temperature above their optimum level will have to migrate of die. Homeothermic animals have a wider option but still may change some migratory habits e.g. coral from the Great Barrier Reef in Australia are migrating south with increased ocean temperature.

10. **Make a list of the probable equipment needed to undertake a scientific monitoring of a nearby waterway. Give the use of each instrument listed.**
This will include the contents of a portable water-testing kit (see Internet sites already given), cleaned collecting vessels, nets, Secchi disks and other monitoring equipment as given in this chapter.

21.7 Reading List

ANZECC & ARMCANZ (2000). *Australian and New Zealand Guidelines for Fresh and Marine Water Quality.* Canberra: Australian and New Zealand Environment & Conservation Council, Agriculture and Resource
Management Council of Australia and New Zealand.

Begon, C. R., Townsend, J. L. & Harper, M. (2005). *Ecology: From Individuals to Ecosystems* (4th Edition). London: Wiley-Blackwell. ISBN: 978-1-4051-1117-1.

Green, R. H. (1979) *Sampling Design and Statistical Methods for Environmental Biologists.* London: John Wiley and Sons. ISBN: 978-0-471-03901-3.

Hill, J. and Wilkinson, C. (2004). *Methods for Ecological Monitoring of Coral Reefs: A Resource for Managers.* Townsville: Australian Institute of Marine Science. ISBN: 0642322376.

Keith, L. H. (Ed.) (1996). *Principles of Environmental Sampling* (2nd ed.). Washington, DC: American Chemical Society. ISBN: 0841231524

Nollet, L. M.L., (Ed.) (2000). *Handbook of Water Analysis.* New York: Marcel Dekker. ISBN: 0-8247-8433-2.

Sheate, W.R. (2009), *Tools, Techniques and Approaches for Sustainability: Collected Writings in Environmental Assessment Policy and Management.* Singapore: World Scientific. ISBN: 978-981-4291-17-0.

Townsend, C.R. (2007). *Ecological Applications: Toward a Sustainable World.* London: Wiley-Blackwell. ISBN: 978-1-405-13698-3.

Chapter 22: Renewable Resources

22.1 Theme

The use of ecosystem services as a supply of food, water, materials and energy resources is more critical now in a world which has been largely separated from these natural ecosystems by urbanisation. Ecosystems as a supply of resources and services can be: provisioning services concerned with the production of food, fresh drinking water, soil and other materials as well as natural forms of energy; regulating services for controlling wastes, diseases, carbon capture or sequestration and promotion of reproduction; supporting services which allow other ecosystem services to operate; and cultural services, such as recreational, aesthetic, cognitive and spiritual benefits to humankind.

22.2 Rationale

There are many important issues which come from the need for renewable resources including the realization that the world currently depends upon non-renewable resources on a vast scale which is going to be difficult, or impossible, to replace. Current use of renewable energy resources is very limited. It will take a long transition time and a change in Human perspective before non-renewable resources are completely replaced.

22.3 Notes on the Practical Work (see PRACTICAL MANUAL 2.)

Experiment 22.1: Sustainability of Tuna as an Ecosystem Resource

The topic of this experiment was chosen because it is a good example of a marine resource which is endangered and involves international competition and cooperation. The Atlantic Bluefin Tuna ((*Thunnus thynnus*) is one of the world's most valuable food resources. There are two distinct populations of Atlantic Bluefins: a western population which spawns in the warm Caribbean and migrates out into the colder central Atlantic Ocean; and an eastern population which spawns in the warm Mediterranean Sea and also migrates out into the Atlantic but keeps closer inshore to Europe and Africa. See:

https://asbtia.com.au/industry/tuna-industry-background/
(Australian Bluefin industry)

https://www.worldwildlife.org/industries/tuna

http://seafoodfrontier.com.au/product/southern-bluefin-tuna/

https://www.washingtonpost.com/world/asia_pacific/tuna-fishing-nations-agree-on-plan-to-replenish-severely-depleted-bluefin-stocks/2017/09/01/7d83c314-8db0-11e7-91d5-ab4e4bb76a3a_story.html?noredirect=on&utm_term=.f2dc0921fc23

This is a simple graphing exercise but it is hoped that the information from the graph and research will show how such a renewable resource can easily be depleted.

PART A: Earlier population numbers

The graph should look like:

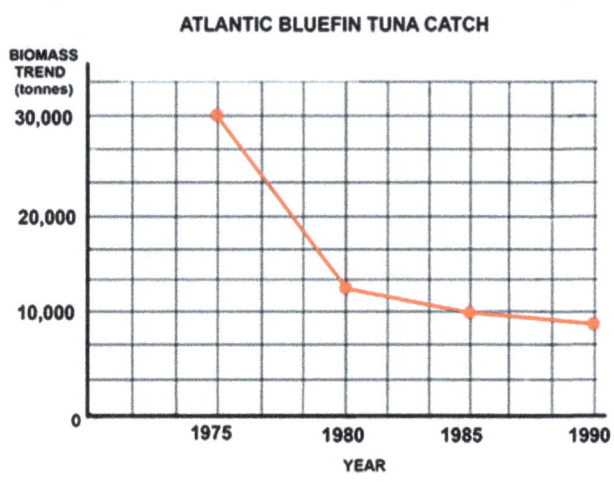

QUESTIONS:

1. What does this graph show about the population of this fish population?
A rapid decline with some levelling out as numbers are depleted and catches become low.

2. Considering that this fish species has been caught commercially in the Atlantic for over 100 years, what would have given this result? (some Internet research in tuna fishing since 1970 would help)
Over-fishing on an international basis.

PART B: Modern times

The graph will look something like:

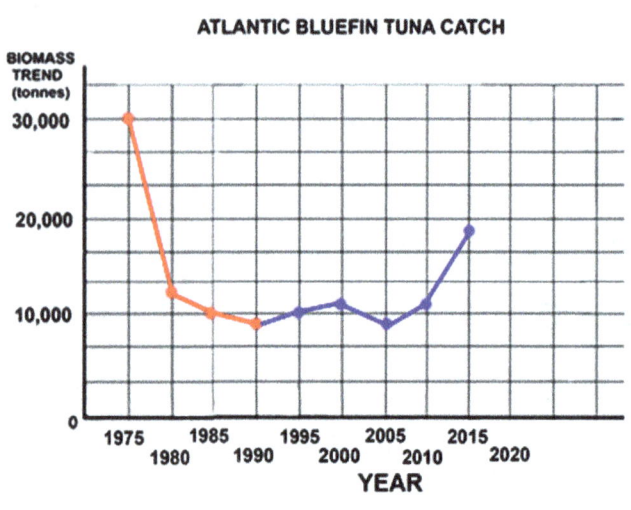

QUESTIONS:

1. What does this graph show about the population of this fish population after 1990?
That the catches have been better – assuming that the population has started to recover if catch sizes are limited to a specified number.

2. What would have given this result? (some Internet research in tuna fishing since 1990 would help).
International agreement to limit catches to specified quota.

3. Is the catching of western Bluefin Tuna now considered sustainable?
There is hope but it would be too early to make such a comment. If the numbers continue to rise to pre-1975 levels then it will be considered as sustainable.

4. Open the following website and read the article carefully. Compare and contrast the data shown in Figure 2 for both population for the years (a) 1974 - 1995 and (b) 1995 to 2015.

https://www.pewtrusts.org/en/research-and-analysis/issue-briefs/2017/10/the-story-ofatlantic-bluefin

If this website does not open try:

https://www.pewtrusts.org/-/media/assets/2017/10/story_of_atlantic_bluefin_tuna_science.pdf

Student group discussion. Answers may be similar to those given above.

5. What would be a factor which could lead to some inaccuracy in these graphs (Figure 2)? Explain.
One assumes that figures given for individual catches are accurate. There would be strong motivation for countries to give false readings and actually bring in catches greater than that stated.

6. What factors caused the sudden decrease in trend of population from the 1970s until the mid-1990s?
Unrestrained fishing by international and local companies with little regard to over-fishing.

7. What were the two main methods of large-scale fishing used in catching tuna? (some research may be needed here to explain these methods).
Long line with individual hooks place along a trailing line and the purse seine method where a net is run in a circular manner around a shoal of fish. There are other methods, see:

https://iss-foundation.org/about-tuna/fishing-methods/

8. What steps have been taken to reverse the overfishing of Atlantic tuna? Is there any cause for optimism that this natural food resource could become sustainable in the future? Why?

Since 1995 there have been international agreements about managing tuna fishing with quotas of catch limits allocated to fishing nations. See:

http://www.fao.org/docrep/012/i1453e/i1453e00.pdf

http://www.fao.org/3/a-y4499e.pdf

https://www.legislation.gov.au/Details/F2016C00642

PART C: Another example -Southern Bluefin Tuna

Depending upon web-browsers used, the pdf file may not show page number progress. Scroll down until two graphs are seen on the one page. Adobe Acrobat is best for this pdf file. Remember that this section is for Southern Bluefin Tuna not the Atlantic Bluefin discussed in previous sections.

QUESTIONS:

1. In figures 1 & 2 give an explanation for the sudden change in catch numbers:

- **(a) After 1960?** - numbers rapidly declined due to unrestricted fishing following a popular demand for tuna.
- **(b) From 1960 to 1990** - continued decline in numbers as catches became fewer in number.
- **(c) From 1990 to 2016** - some increase in numbers due to international regulation but numbers have now levelled out at a low value.

2. What changes occurred to the methods of fishing after 1999?
Not detailed in this document but there was a change in fishing methods with purse seine fishing using fish aggregating devices and more small-scale long line fishing. See:

http://www.fao.org/docrep/007/y5428e/y5428e03.htm

3. Considering the catches over the last 20 years do the graphs show any hope for sustainability of Southern Bluefin Tuna in the future? Discuss.
Possible. Catches and stocks seem to be maintained at a low level with some hope for the future with international management.

4. Scroll down to Figure 4. What do these images show about changes in the geographical locations of tuna fishing catches? (a series of 5 maps)
Numbers have decreased but there still is a high concentration of catches in southern Australian and New Zealand waters.

5. What attempts have been made to reduce over-fishing and manage Southern Bluefin Tuna as an environmental resource?
Refer to the following websites:

The following websites may assist in these questions:

https://theconversation.com/australian-endangered-species-southern-bluefin-tuna-11636

https://www.dpi.nsw.gov.au/__data/assets/pdf_file/0004/508018/southern_bluefin_tuna_sis_part_1.pdf

https://www.ccsbt.org/en/content/about-southern-bluefin-tuna

CONCLUSIONS:

Write a report about the use of tuna as an ecosystem resource. Comment on demand, uses, fishing catch trends from the 1960's with reasons for these trends, methods of fishing and attempts to make the industry sustainable.
This will be a general summary of what has been gleaned from the above graphs and websites. It could follow an extensive student discussion.

Experiment 22.2: Sustainability of Timber as an Ecosystem Resource

This is a good experiment in the ability to search for, record and collate data from a complex data base but it may take some time. Some assistance may be needed to show the students how to operate the site e.g. the PLANTATIONS button must be found at the bottom of the LAND COVER tab and turned on to get this data. Make sure to click on the NAME of the country to call up the ANALYSIS operation. Some countries such as Australia are not listed for plantations so see:

https://www.carbonbrief.org/mapped-where-afforestation-is-taking-place-around-the-world

This site gives values in hectares so students will have to calculate this as a percentage value from the total land area in Mha from the original site.

RESULTS:

Draw a graph of the gathered data (note other resources below may have to be used to complete the table about plantations) e.g.
This graph is probably best drawn as a column graph for each country for some of the values obtained e.g. percentage tree cover with percentage plantation cover. Data for the countries given is provided below but it would be interesting to also look at Russia, Canada and perhaps as a contrast another desert or European country.

COUNTRY	TOTAL TREE COVER (%)	TREE COVER LOSS (MHa)	TREE COVER GAIN (MHa)	TOTAL TREE COVER LOSS (%)	PLANTATIONS %	COMMENTS Plantations areas in ha.
Australia	32.0	4.2	1.41	9.9	0.26	2017000
Indonesia	60	24.4	6.96	15.0	2.6	4946000
Brazil	47	50.9	0.89	9.8	0.9	7736000
China	13	0.05	2.24	5.4	0.4	78982000
New Zealand	35	1.08	2.64	9.6	7.5	2087000
United States	23	36.1	13.8	13.0	2.74	26364000

Also see: https://farm1.staticflickr.com/300/31552050673_1167977beb.jpg
(map showing changes in forest plantation growth by country)

http://www.fao.org/docrep/w4345e/w4345e03.htm
(scroll down to details about plantations Tables 2 & 3)

VIDEOS:

1. Go on a mini-excursion to Puerto Maldonado, a logging and gold-mining town in the Peruvian Amazonia, with the author and his wife in 2011 and travel down the Rio Madre de Dios (River of the Mother of God) and into the rainforest at:

> https://www.youtube.com/watch?v=VhJ7Ve1FbL0
> (38 minutes)

2. State of the world's forests with special reference to the Amazon from GRIDArendal - a Norwegian foundation working closely with the United Nations Environment organisation.

> http://www.grida.no/resources/8444
> (3 minutes)

QUESTIONS:

1. What factors must be kept in mind when looking at the total forest cover of any country?
Many countries such as Australia have large, natural deserts which are not considered when looking at timber as a resource.

2. Which of the selected countries has the highest (a) forest lost and (b) forest gained (c) amounts of plantation?

> (a) Brazil, the United States and Indonesia are the top three.
> (b) New Zealand is the ONLY country with overall gain. This is a good example of selective use of statistics. The website data says that New Zealand has a 9.6% loss of total tree cover yet the data in MHa shows a

gain. One can only assume that the gain is due to planted tree cover and that the 9.6% loss is that of natural tree cover since 2001 and does not include any re-planting.
(c) China and New Zealand

3. Convert the plantation values (MHa) in Question 2(c) to percentages of total forests and rank the listed countries in order of plantation percentages.
See the Table given above for percentages and absolute values

4. Looking at the graphs of tree cover loss (from the Comments), which of the listed counties has the biggest problem with deforestation?
Brazil with 50.9 Mha lost but only 0.89 gained.

5. Not all of the values for deforestation are due to timber being logged as a resource. List other reasons for deforestation (some research needed).
These include natural losses due to disease and climate change.

6. How does deforestation affect climate change?
Removes the natural transpiration water loss from the trees and exposes the ground which then has a different albedo and also heats the air above, drying out the region.

7. "Plantations are not true forests." Discuss this statement from an ecological viewpoint.
Like many artificial features, plantations do not usually represent nor are part of the original tree ecosystem. Planation forests are often introduced species.

8. Are there any problems associated with timber plantations?
see: http://www.forestnetwork.net/Docs/Gippy_h2o.htm#plantations

9. What timbers are used worldwide in plantations and as fast-growing trees for reforestation? Indicate some case studies where these timbers are being used.
Many fast-growing timbers are used such as radiata pine and some eucalypts e.g. eucalypts in some South American countries.

10. What are some strategies in restoring the world's forests and using timber as a sustainable ecosystem resource?
See also:
http://www.global-economic-symposium.org/knowledgebase/the-globalenvironment/protecting-and-restoring-the-world2019s-forests/figure_overlay

Changes in world biomass
http://www.fao.org/docrep/003/y0900e/y0900e05.htm
(scroll down to Figure 8)

http://www.forestnetwork.net/Docs/AlternativeVision.htm
(Future alternatives to logging)

CONCLUSIONS:

Write a short report on the current state of the world's forest resources, causes of depletion and strategies being used to restore native forests and use timber as a sustainable resource. Include any diagrams, maps or photos as required giving appropriate references and acknowledgements.
This will come from an overall answer to the questions above and may follow a group discussion.

RESEARCH: (Optional)

Use the Internet to find out how renewable timbers can replace current building and other materials which use non-renewable resources as their raw material.
This might include the innovative use of bamboo and laminated timbers and fibres. See:

https://www.dw.com/en/wood-renewable-construction-material-of-the-future/a-42012053

https://www.woodproducts.fi/content/wood-a-renewable-natural-resource

https://www.swedishwood.com/about_wood/choosing-wood/wood-and-the-environment/wood-is-a-sustainable-construction-material/

Experiment 22.3: Sustainability of Surface Water as an Ecosystem Resource

This is another detailed data analysis using secondary data from websites. Students should read the information on these sites carefully and make any summary notes as appropriate. It may take some time and students should either open several sites together on a PC or share tablets each with different sites. The first website is of general use but the second website is more specific and useful.

QUESTIONS:

1. What are major sources of the World's freshwater and their relative percentages? (See Sites 1 & 3)
Site 3 will take some time to load. Good data is also available from resources in the textbook. Site 3 also had difficulties opening in Chrome so Firefox was used with success. Slide 3 shows the total hydrosphere with glaciers, groundwater and surface waters (in order) as less than 2.5 % of total.

2. How are Australia and the United States generally assessed for water scarcity (Figure 1 Site 2 and Slide 5 Site 3 - Note: SEI index has red as critical)? Comment on the main regions of these countries affected.
Australia and central United States have high scarcity in the red zone.

3. What countries are regarded as uniformly very critical when it comes to water scarcity? (see references above especially Slide 5 Site 3.)

See Slide 5 site 3 and look for the red zones.

4. Comment on the use of groundwater (i.e. underground water - artesian and sub-artesian) as support for groundwater in irrigation and stock raising (Site 2, p 13 & 14. Note: Abstraction refers to extraction of groundwater).
The first map shows crops using groundwater and the second the increase in abstraction. Australia has had only a modest increase but some countries such as India, eastern China and Eastern Europe have had significant increases.

5. What is a major problem with contamination of surface water? (Site 2, pp. 14-17).
Waste water containing pathogens and agricultural runoffs with fertilizers and pesticides are considered major problems.

6. What will be one of the main problems with soil moisture due to future global warming? (Site 2, Figure 6, p.18).
Soil moisture content will be considerably reduced with increased temperature – probably assisted by deforestation.

7. What are some ways in which water can be conserved? (Site 1, Site 2, Figure 1.4, p.31, Table 1.2., p.32).
There are a great number listed in Table 1.2, including reforestation, water harvesting, reduction of concrete and pavements and adding green spaces in cities.

8. What will be some of the ecosystem services which will be used in improving water use and sustainability? (Site 2, Table 1.1., p. 30).
Also, a long list including use of freshwater for drinking, soil stabilization, reduction of water hazards, influencing local precipitation and providing hydropower.

CONCLUSIONS:

Write a general assessment in several paragraphs concerning the availability of water in the future, the types of demands, both natural and man-made which will impact on available water sources and some of the possible ways by which water will be conserved and used as a sustainable resource. In your assessment use diagrams and data as required to support the content.
From a general group discussion. The tables from site 2 are good for this summary.

RESEARCH: (Optional)

1. Continue with the examination of website 3 (Slide 8 onward) and note the major current and future problems with developing countries such as those in Africa. What are the problems and what political tensions are likely to develop over water supply?   or

2. Use Site 2 to examine more closely the problems with ground (sub-surface) water. Some specific Internet research may be required with specific reference to Australia or the United States.

This will depend upon student preference and research. It might be useful as a group forum, especially the socio-political problems with countries sharing limited water resources such as Sudan and Egypt.

Experiment 22.4: Case Studies – Surface Water as a Sustainable Resource

This is an important experiment because both of these major river systems are vital for their nations and also under threat. Recent fish extinctions in the Murray-Darling system suggests that a combination of natural water loss and poor management has resulted in an environmental disaster which may not have an easy nor short-term solution.

Each part uses a question-and-answer technique as a comprehension exercise of the information given in the websites. One or both case studies can be attempted using group analysis and discussion. This may take the form of a major group forum about one or both of the river systems with general notes, discussion and debate replacing the traditional student Practical Report.

Experiment 22.5: Geothermal Energy as a Sustainable Resource

This is another good data base search and analysis on geothermal energy. Whilst most students will think of the obvious countries using such energy sources today, notably Iceland, New Zealand and the United States, it is surprising to note the extent of geothermal use around the world. In addition, there is also some potential in most countries to use deep heat sources in crystalline rock which is often heated by radioisotopes present in the rock.

Website 3 will give an interesting interactive map but students will need to scroll using the left mouse and then click on the country to get data or type the country's name into the search box. Further details from the pie graph presented can be had by clicking on the sections of legend or the pie graph for each energy.

COUNTRY	ENERGIES GIVEN as a PERCENTAGE						COMMENTS
	GEOTHERM.	COAL	NATURAL GAS	NUCLEAR	HYDRO	SOLAR/ WIND/ etc.	
WORLD	0.05	28	21.6	4.9	2.4	0.9	Oil is top 32%
AUSTRALIA	0.02	78.3	14.8	0	0.3	0.5	Coal is top
UNITED STATES	0.4	21.4	31.5	10.7	1.1	1.1	Natural gas is top
NEW ZEALAND	29.4	11.7	24.4	0	12.8	1.5	Natural gas & Geothermal
ICELAND	75.8	0	0	0	24.1	0	Very renewable
SWEDEN	0	0.3 peat	0	43.8	19.1	4.2	Nuclear is top
JAPAN	7.9	0 ?	7.8	8.1	24.2	12.7	37.6% biofuels
CHINA	0.2	74.9	4.5	1.8	3.8	1.6	Coal is top
TURKEY	15.3	40.4	1.0	0	18.2	6.3	10.4% biofuels
INDONESIA	4.1	57.4	15.4	0	0.3	0	Coal is top
RUSSIA	0	15.0	39.3	3.8	1.1	0	
OTHER							

This is an interesting graph showing how little some countries use some of the traditional non-renewable energies. Some countries have a singular dependence on one energy e.g. Australia and China for coal and Sweden (and probably other western European countries) for nuclear.

1. What is the difference between the use of geothermal power and geothermal heat? How much of each were used in 2014-2015? (Site 1).
Geothermal energy comes from natural steam from the Earth whereas geothermal heat requires water to be pumped into hot rocks. See the graph.

2. On this site and others, energy units are in often in PJ and Mtoe. What are these units? power is also given in GW. What is this unit? (extra research needed).
PJ stands for petajoule (NOT to be confused with the very small unit picojoule pJ) which is 10^{15} joules. Mtoe stands for million tonnes of oil equivalent and is a unit of energy defined as the amount of energy released by burning one tonne of crude oil and is approximately 42 gigajoules (10^9 J).

3. What is the difference between energy and power? (research).
Power (in watts or gigawatts GW or 10^9 W) is the amount of energy (in Joules) used per second.

4. Look at the graph at Site 2 which shows the depletion of flow rate from an operational hydrothermal well. Use the graph to find out when (year) the well will have zero flow.
A tricky question as the graph is logarithmic. A simple way is to make multiple copies of the graph and lay them together. Estimation by this method (i.e. graphically) will have the well depleted in about 2019-2020.

5. From the table constructed of geothermal and other energies for selected countries:

>**a. Comment on the world use of geothermal energy, giving its percentage use and rank compared to the other energies** - from the table it can be seen that most countries have little or no use of geothermal energy. It is those countries with natural volcanic sources of steam which make use of such power e.g. Iceland (76%) and New Zealand (29%).
>**b. rate the selected countries in order of their use of geothermal** - Iceland, New Zealand, Turkey, Japan, Indonesia etc.
>**c. Of those countries having high geothermal use, comment on their use of other renewable energies (some additional research may be needed).**
>They also make good use of hydroelectricity although New Zealand also uses natural gas. Japan seems to have a good mixture of several energies.

6. What are the factors which control the sustainability and renewability of geothermal energy?
The main factor is the pressure of the natural geothermal steam in those countries using this resource. With the use of geothermal heat it will depend upon the cooling rate of the heat source once water is pumped down. This will depend upon the size of the rock body and its natural re-heating. Other problems associated

with this system is the contamination, especially radioactivity which needs to be avoided by using heat exchangers at the surface to produce the hot fluid to drive turbines.

CONCLUSIONS:

Write a brief report in paragraph form explaining:
   a. why geothermal energy is considered a renewable and sustainable form of energy;
   b. what are the necessary sources of geothermal power and the different types available;
   c. What are the limitations of geothermal energy; and
   d. What is its current use and potential for the future?

Again, this will depend upon the students' ability to use the information provided and it may be useful to have a group discussion about the conclusions.

RESEARCH: (Optional)

1. Use the internet to find the location of major geothermal power stations.
2. Distinguish between hydrothermal and hot dry rock (HDR) power systems. Use a simple diagram to explain how each generates power from the Earth's heat.

See the text book and Figure 2 of website 3.

Experiment 22.6: Porosity and Permeability

This is a relatively simple experiment to the show the difference between these two soil/rock parameters and their importance when studying groundwater.

POROSITY is the amount of pore space (open space) between the grains of sediment. It is found by:

$$\text{POROSITY \%} = \frac{\text{Volume of pore space}}{\text{Total volume}} \times \frac{100}{1}$$

PERMEABILITY is the rate at which water moves through sediment and this depends upon several factors including the nature of the sediment, the cross-sectional area of the layer of rock and the height of water intake over outflow or the head of water.

The methods for both parts are straight forward and make use of simple laboratory equipment. Dry sand or silt or gravel can be used for the medium but they must be dry and relatively homogeneous. Rough material may have to be sieved to get a uniform medium and the sieves will also give an indication of size of the particles which can be checked against sediment data for different Wentworth Scale sediments and their porosity/permeability values.

QUESTIONS:

1. Why are hydrologists and other people interested in the porosity and permeability of soil, sediment and rock?
Hydrologists are most interested in these values so that they can predict volumes of reservoirs in groundwater aquifers and also to determine rates of recharge by knowing the permeability of rock between the aquifers and the recharge areas.

2. Explain how a rock could be porous but not permeable? Why would this matter to exploration geologists looking for water or oil?
Rocks can be porous but if the pore spaces are not connected then there will be no permeability of flow. Sub-surface fluids need to seep into containing porous rock from their source of organic decomposition (oil & gas) or from their recharge area.

3. What is the purpose of the paper wad in the permeability experiment?
It simply stops the soil medium from falling into the flask. To ensure that one is measuring the permeability of the sediment and not the wad, a time for the permeability for the wad alone is a good control.

4. What is the purpose of the control experiment?
To ensure that it is the permeability of the sediment that is being measured.

CONCLUSIONS:

1. List the sediments in order from best to least for (a) porosity and (b) permeability:
(a) Porosity of surface soil typically decreases as particle size increases so the silt will be more porous than the sand and then the gravel. The smaller sized particles allow for many more small voids between them.
(b) Permeability depends upon the joins between these voids and because they are bigger in gravels, they are more likely to allow water etc. to pass from one to another so gravel is more pervious than sand which is more pervious than silt.

2. Are these lists the same? Explain any differences or similarities.
No. They have an inverse relationship – see above.

3. Which of the permeability methods (dry or wet) would be a more honest measure of a sediment's true permeability? Explain.
Having wet sediment in the permeability experiments ensures that any added water will not be retained by the sediment in wetting (i.e. coating and staying) the dry sediment.

4. What could reduce the porosity of a sediment in nature?
In nature, porosity can be reduced by having a finer matrix filling the voids if the original sediment is poorly sorted or if there has been some compaction or metamorphic action on the rock.

5. Comment on any errors which could occur in the measurements.
The main errors here are in having the porous sediment settled in the beaker so that it is not too loose nor compacted. The beaker of sediment should be gently

tapped as the sediment is added. Also the estimation of the water level in the sediment as the water is slowly added may be an error if air bubbles form within the sediment. In the permeability experiment, the wad must be able to stop sediment falling into the flask but not too impervious itself. It also should be wet and a control trial should be done without sediment.

RESEARCH: (Optional)

How can rock have its porosity and/or permeability changed? How are these rock parameters useful in engineering of roads, dams, bridges and building construction?

AS mentioned, these parameters can change due to compression (weight of material on top or Earth forces) and in-filling with a finer matrix etc. Engineers need to know about these parameters because the flow of water through soil and rock are important for the water integrity of dams. These parameters and their water content can also change the strength of rock, an important consideration in foundations of buildings, road construction and side banks and benches.

22.4 Other Activities

- Make some simple rectangular structures out of bamboo and see how much weight they can hold or hang weights on the ends of different wood slats of the same length, including bamboo and see which is the strongest. Research the use of bamboo in Asia as a scaffolding.

- Research the use of distillation plants in obtaining drinking water from the sea, especially in desert countries.

- Make a simple solar distillation plant to obtain freshwater from saltwater. See:

    http://www.appropedia.org/Solar_distillation

    http://all-about-water-filters.com/ultimate-guide-to-solar-water-distillation/

    http://www.sunfrost.com/blog/producing-distilled-water-for-batteries/

    https://www.researchgate.net/figure/A-schematic-diagram-of-a-simple-solar-still_fig1_294577641

- Use spherical plastic beads or marble to test porosity of different sized beads or use beads of the same size but different shapes to test porosity and permeability.

- Set up a simple hydroponic system in a warm, sunny window using a trough garden pot, sterilized sand or ceramic beads and use a complete plant food from the store. Water every two days as required or have a trickle system with small pump to replace water in the tank above. 100 ml diameter PVC pipe with circular holes cut every 10 cm and filled with gravel then turned on its side makes a good trough but seal the ends.

22.5 Answers to Multichoice Questions

Q1.A   Q2.D   Q3.B   Q4.B   Q5.C   Q6.B   Q7.A   Q8.B   Q9.C   Q10.B

22.6 Some Suggestions for the Review Questions

1. **What are the major issues facing the next generation? What changes in lifestyle will the average city-dweller need to make to live in a sustainable world?**
   This is going to be a big social challenge to the current and next generation. Electrical power may not be cheap initially as renewable systems take over from the traditional fossil fuels. Housing will have to be designed with the environment not against it with the mass use of air-conditioning being restricted. Open flow ventilation and the use of renewable building materials will have to be considered. Many of the older practices still used in some developing countries such as use of mud-brick houses, timber construction and natural water and plant use for interiors. Recycling of materials such as metals, glass and paper will become more important and other organic wastes will be used as a source of energy and fuels. Transportation will need to be less personal with public transport using renewable energies replacing individual vehicles moving into city centres.

2. **How can natural ecosystems regulate the environment? Give an example of such regulation in:**

   (a) **the natural environment** – trees purify the air with increased oxygen and water vapour and remove carbon dioxide.
   (b) **the urban environment** – similarly additional green spaces within cities and buildings will assist in refreshing urban air quality.

3. **Many of the natural resources used today are often over-harvested. Give two examples of such over-harvesting and suggest possible management practices which could prevent it.**
   This concerns natural resources and not those grown or grazed. Over-harvesting of fish, water and natural timbers are good examples of natural resources now being seen as scarce.

4. **What is bamboo? What could it be used for to replace non-renewable resources in a modern society?**

Bamboo is a quick-growing tall grass that is cylindrical in shape and strengthened along its length so it is very strong. It can be used as a stand-alone building material, as it has been for thousands of years in Asia or used in building construction as scaffolding and other uses. It can also be sliced, made into fibre, stripped into threads and woven and as shoots can be eaten.

5. **Thinking about other renewable materials which could be used to replace current non-renewable materials, research the following:**

    (a) **Making biodegradable/recyclable plastic** – these can be made from natural products such as corn starch. See:
    http://www.pepctplastics.com/resources/connecticut-plastics-learning-center/biodegradable-plastics/

    https://theconversation.com/why-compostable-plastics-may-be-no-better-for-the-environment-100016

    (b) **Large-scale hydroponics** – very useful at the home level growing herbs and vegetables in a home-made nutrient of plant food from the store in beads or sterilize sand in a sunny position indoors during winter. It can also be done on a large scale with water circulation in large greenhouses and controlled nutrient. See:

    https://www.fullbloomhydroponics.net/hydroponic-systems-101/

    https://www.hydroponicxpress.com.au/learn-hydroponics/

    (c) **Natural fibres to replace synthetics in clothing** – synthetics are a product of the oil and coal industry from the 1940s. They often replace good natural fibres such as cotton, wool and silk.

    (d) **Alternative building materials** – again some of the old ideas were good. Use of timber frames, recycled glass (e.g. bottles), mud-brick, bamboo etc. with a change in design needed to match the local environment,

6. **Research the various methods of disposing of waste materials. What are some zero-emission alternatives?**
    Several systems already being used in several countries. Some rely on high-temperature incineration or furnaces which generate heat from burning waste put there is still a problem with emissions. See:
    http://www.grrn.org/page/zero-waste-and-climate-change

http://www.fujitsu.com/id/Images/2002report29_30_e.pdf

https://www.panasonic.com/global/corporate/sustainability/eco/resource/zero.html

7. **Do a renewable/waste material audit of (a) the home and (b) the workplace to find what non-renewable materials could be replaced by renewable materials or recycled materials.**
This could be an interesting exercise if done as a group effort. There is now some motivation to recycle metal. Glass and cardboard containers for cash. It may be a useful exercise in finding the location of a wholesale recycling company and then organising the collection of glass bottles and cans in the school or college and then selling it.

8. **Use the Internet to find out the ecological footprint of the average person in an urbanised, western community. Thinking of local and personal lifestyles, how could that footprint be reduced?**
Try an 'ecological footprint calculator' – it might not be scientifically accurate but it does demonstrate the idea. Go to:

http://ecologicalfootprint.com/

https://www.earthday.org/take-action/footprint-calculator/

9. **Many countries such as Australia and the United States often have problems with drought. Research how many ancient civilizations living in arid countries coped with drought situations. What could be some long-term solutions to drought-proofing places such as inland Queensland or the southwestern United States?**
A variety of research ideas here. Look at some of the traditional ways that some desert-dwelling civilizations constructed water systems and dwellings. Take a critical look at local use of water collection and storage and also building design and the environment.

10. **"Environmental action and conservation are a matter for governments – there is little that the average person can do." Comment on this statement by preparing for a group discussion.**
The average person can do a lot for the environment by looking at their own use of resources and adopting some conservative habits, especially those mentioned in this book. Governments are usually reluctant to change because this loses votes. Environmental activism if extreme also loses support. Future social change as well as political attitude change is also needed.

22.7 Reading List

Akitsu, T. (Edit.) (2018). *Environmental Science - Society, Nature, and Technology.* Singapore: Pan Stafford. ISBN: 9789814774963.

Daniek, M. (2013). *Do It Yourself 12 Volt Solar Power (Simple Living).* East Meon, Hants, United Kingdom: Hyden House Ltd. ISBN: 9781856230728.

DeGunther, R. (2009). *Solar Power Your Home for Dummies*, 2nd Edition. New York: John Wiley & Sons. ISBN: 9780470596784

Rao, K.R. (2018). *Wind Energy for Power Generation - Meeting the Challenge of Practical Implementation.* Basel, Switzerland: Springer International Publishing AG. ISBN: 3319751328.

Wright, DR. J., Osman, Dr. P. & Ashworth, P. (2009). *The CSIRO Home Energy Saving Handbook - How to Save Energy, Save Money and Reduce Your Carbon Footprint.* Sydney: Pan Macmillan Australia. ISBN: 9781405039611.

Chapter 23: Renewable Energies

23.1 Theme

Renewable fuels offer some hope for the future because they have less impact on the environment and provide the security of future energy availability. They include hydro-electricity, solar power, wind power, tidal power, biogas, ethanol and other biofuels, geothermal power, and the use of hydrogen as a fuel.

23.2 Rationale

Currently, the world faces an energy problem, with an increasing demand for more energy conflicting with the problems caused by the burning of fossil fuels and the subsequent problems of global warming. Energy will be in great demand in the future, but there is a need to prevent any further rise in atmospheric and oceanic temperatures.

23.3 Notes on the Practical Work (see PRACTICAL MANUAL 2.)

Experiment 23.1: Hydroelectricity and Wind Power

This simple experiment uses basic DIY or toy devices to show the principle of water and wind power. Most students within a modern, urbanized society may not be aware that these sources of power were probably the original drivers of the pre-industrial revolution. The experiments attempt to show the principles of both sources and that unsophisticated systems can be built to generate electricity. Toy versions of these can be purchased in scientific/toy shops if one does not want to build a DIY model.

For PART C, scroll down to the interactive section on energies by countries. The table will be similar to that given in a previous experiment on geothermal energy:

COUNTRY	MAJOR SOURCE of ENERGY %	HYDRO	SOLAR/ WIND/ etc.
WORLD	Oil 32 Coal 28	2.4	0.9
AUSTRALIA	Coal 78.3 Nat. Gas 14.8	0.3	0.5
UNITED STATES	Nat. Gas 31.5 Oil 28.8 Coal 21.4	1.1	1.1
JAPAN	Biofuels 37.6	24.2	12.7
CHINA	Coal 74.9	3.8	1.6
INDIA	Coal 47.5	2.1	0.9

Again, this site has some problems with Chrome but Firefox opened eventually.

QUESTIONS:

1. What is the relationship between fluid (water or air) flow rate and the amount of electricity generated?
Generated electricity is directly proportional to flow rate.

2. Is it the amount of fluid or its flow rate which has the greatest effect?
Flow rate

3. How could this be regulated in hydroelectrical power stations?
By valves in the pipelines to the turbines.

4. What are the main requirements for:
- a. **hydroelectrical power** - constant supply of water at a higher altitude i.e. good rainfall and mountain dams.
- b. **wind power farms?** – open spaces and a constant wind flow, preferably from a constant direction.

5. How can the energy of wind farms be stored when the wind drops?
This can be done through the use of:

- Battery storage
- Compressed air storage
- Hydrogen fuel cells
- Pumped water storage.

6. What is pumped hydroelectric energy storage (PHES)? Why is it used? Give an example of a location of one of these systems.
Water is pumped up hill into a storage dam using electrical pumps when there is spare electricity so that it can be allowed to flow downhill and generate more electricity during peak times e.g. the proposed Snowy Mountains 2 project and the Splityard Creek Dam which takes water from Wivenhoe Dam in Queensland and then returns it via a hydroelectricity power station. There are over 240 stations in the world with great potential. See:

https://www.csenergy.com.au/what-we-do/generating-energy/wivenhoe-power-station

https://theconversation.com/want-energy-storage-here-are-22-000-sites-for-pumped-hydro-across-australia-84275

CONCLUSIONS:

1. Comment on the structure and processes required to make energy by these two methods.
Hydroelectricity is a well-established source of energy and many countries have been using it for a long time. Wind turbines are not new but they are only now being used on a major scale. Hydro relies on a good rainfall consistently in the catchment areas and must use dams built in mountainous regions. Global warming

and the threat of less rain in some parts of the world may be a threat to some hydro schemes. Wind power also relies on a constant source of wind from a prevailing direction and are thought by some to be environmentally unfriendly visually.
See:
https://water.usgs.gov/edu/wuhy.html

2. Discuss the advantages and disadvantages of hydroelectricity and wind power generation.
See above.

3. From the table, what is the current importance of both of these forms of energy generation compared to other forms currently used and is there any country which seems to be using these forms of power more than others and why?
Currently hydro and wind power are only a small fraction of the energy generated by fossil fuels. Some smaller countries with abundant water such as Albania, Paraguay, the Democratic Republic of the Congo and Nepal have almost all of their energy needs met through hydroelectricity. The main production of hydroelectricity is in countries with good water but high energy needs such as China, Brazil, Canada and the United States. There are no countries which are able to rely only on wind power, but the biggest producers of this energy are China, the United States, Germany and India.

4. What is the likelihood that these two forms of energy will play a more important role in energy production in the future?
See:
https://ourworldindata.org/renewables)

https://www.nytimes.com/2017/02/09/business/energy-environment/wind-energyrenewable.html

RESEARCH: (Optional)

Use the Internet to find out the location of some of the world's largest wind farms and find out how they generate electricity.

See:

https://interestingengineering.com/the-11-biggest-wind-farms-and-wind-power-constructions-that-reduce-carbon-footprint

https://www.worldatlas.com/articles/the-10-largest-wind-farms.html

Visit a wind turbine and see inside at:

https://www.energy.gov/eere/amo/mining-industry-profile

Experiment 23.2: Solar Power

The crucial part of this experiment is having a good set of solar cells. The experiment uses two cells recycled from solar lighting modules which worked very well.

PART A is simply an exploration of the output of solar cells. This will depend upon the cells used but generally they will be low voltage and DC current. Like any cell, when placed in series the voltage and the current is added and in parallel, the voltage remains the same but the current is doubled. This assumes that the solar cells are in full sunlight. The output drops quickly when in shadow.

QUESTIONS:

1. What factors would give a reduced value for the readings from the photocell?
Reduced lighting and poor connections.

2. What are the instrument errors of the (a) voltmeter and (b) ammeter? (remember the rules about analogue and digital scales).
Depends upon the instrument used. For analogue meters it will be half of the smallest unit and for digital meters one smallest unit.

3. What is the difference between a series and parallel connection in an electric circuit?
Series circuits are connected positive to negative etc. which adds the voltage and the current of each cell. Parallel circuits have positive to positive and negative to negative giving the same voltage but more current with the safety factor that if one cell fails, the circuit still works,

PART B: The efficiency of photovoltaic cells

QUESTIONS:

1. What is the efficiency of the single cell(s) used?
This will depend upon the cell(s) used but solar cells are rarely above 20% efficient.

2. What are some of the efficiencies of some common solar panels?

See:
https://news.energysage.com/what-are-the-most-efficient-solar-panels-on-themarket/

3. What are some of the factors which also determine the efficiency of a complete solar panel system?

See: https://solarcalculator.com.au/solar-panel-efficiency/

CONCLUSIONS:

Generally, comment on the pros and cons of solar cells and how much power they can produce. Also comment on the method used and the errors involved.
Solar cells connected in series and parallel can develop modest power outputs for the domestic home and there are some solar farms which give significant output. Initially the establishment of such extensive arrays of panels can be high due to manufacture and set up costs but eventually they provide a satisfactory output in strong sunlight. There has been some criticism that such arrays have been built upon good agricultural land and the future will depend upon these being built on land that is non-productive. Costs of panels must also include the use of non-renewable materials and a considerable amount of energy to manufacture the panels, but it has been found that a solar panel system generally pays for itself after about eight years.

See:

http://www.dani2989.com/matiere1/solar0710gb.htm

1. What was the potential difference and current output of the single cell in full sunlight?
2. What was the efficiency of the single cell?
These questions depend upon the cells used.

3. What was the effect on adding the cells together in (a) series and (b) parallel?
See answers given above about series and parallel connections.

4. **What is Ohm's Law? Does this apply to the connection of voltaic cells? Explain.** Ohm's law states that the current (I) through a conductor between two points is directly proportional to the voltage (V) across the two points. Introducing the constant of proportionality, the resistance, one arrives at the usual mathematical equation that describes this relationship:

$$V = I \times R$$

EXERCISE:

A typical solar panel system may be rated at 6.6 kW and have up to 22 individual panels of 300 W each. It may also be attached to one or more lithium ion battery packs for night use and an inverter for changing the direct current input to 240 v (110V in the US) alternating current.

Consider a typical home using only solar power as given above at meal time at night. The family may have switched on: a refrigerator (20 cf or cubic foot) and freezer (14cf) which are permanently on, a stove with two hotplates, an electric wall clock, 4 incandescent light bulbs at 100 W each, 1000 W microwave and a plasma TV.

Question: will the home solar panel system and battery allow all of this to happen?

see:
https://coolaustralia.org/wp-content/uploads/2013/12/Typical-power-ratings-forappliances.pdf

Note that 1 cf = 28.3 L, so the 20 cf refrigerator is equivalent to 560 L and the 14 cf freezer is about 396L. These values will vary depending upon the efficiency of the unit.

See also:

http://www.ecospecifier.com.au/media/146119/100805_pv_appliance_load_calculator.pdf

https://www.e-education.psu.edu/egee102/node/1915

https://www.cpp.edu/~pbsiegel/sci210/appliances.pdf

The total power used by the family for (say 1 hour at meal time) could be:

Refrigerator   (say)   1000w
Freezer   750
2 hotplates   2200
Wall clock   (say)   10
4 lights   400
Microwave   1000 less for short use.
Plasma TV   300

TOTAL: = 5.66 kw used over 1 hour so energy is 5.66 kJ

Assuming that the output of the solar panel and battery is 6.6 kw then this amount of power would be able to be used by the family.

RESEARCH: (Optional)
Use the Internet to find out how much energy and raw materials are needed to produce one 300 W panel. Are there any environmental problems?

See:

http://astro1.panet.utoledo.edu/~relling2/PDF/pubs/life_cycle_assesment_ellingson_apul_(2015)_ren_and_sustain._energy_revs.pdf

https://www.lowtechmagazine.com/2008/03/the-ugly-side-o.html

Experiment 23.3: Hydrogen Gas as a Fuel Derived from Solar Power

Again, this experiment relies on the power of the solar cell to dissociate water (with a little sulfuric acid) to give hydrogen and oxygen by electrolysis. It worked well after a relatively short time with the cells shown. Several cells may have to be connected in series to get a better result. It also may be a good idea to prepare a few tubes of oxygen gas and hydrogen gas using ordinary power transformers and an electrolysis apparatus to show the effects of a flame on each gas.

QUESTIONS:

1. Why must both test-tubes be completely filled with water?
So that any gas produced is not contaminated with air.

2. Why was the sulfuric acid added?
To assist in the conduction of the electricity,

3. Which test-tube will produce the hydrogen gas? Explain.
The hydrogen gas will be formed in the test-tube over the negative terminal as the positive hydrogen ions in solution are attracted to it.

4. How can one observe that the water is splitting up by the electricity?
Bubbles will be seen in each tube with two volumes of bubbles of hydrogen over the negative terminal and one volume of oxygen over the positive terminal.

5. Why should one be very quick in performing the test with the lighted match or taper?
The hydrogen gas is very light and will quickly leave the tube.

CONCLUSIONS:

1. What is produced by the electrolysis of water?
Hydrogen and oxygen gases in the ratio of 2:1 by volume.

2. Was the use of the voltaic cells practical? If not, how could this system be improved?
It worked well. Any improvement would be the addition of more cells for greater electrical output.

3. Is hydrogen gas as a fuel a good alternative to the current use of petroleum products? Why?
It has good potential but still requires energy to make it and to compress it. Using renewable energies such as solar and wind to electrolyse water would be a good application. Currently there are still some problems with supply and the social mistrust of hydrogen.

4. What waste product is produced by (a) electrolysis of water and (b) combustion of hydrogen gas fuel?

Electrolysis of water produces hydrogen and oxygen gas. When hydrogen is burnt in an internal combustion engine with air it will give steam as the only waste gas. Hydrogen and oxygen can also be used in fuel cells with the same waste.

RESEARCH: (Optional)

Use the Internet to find out how hydrogen gas is currently used in powering vehicles.
See:

https://www.eia.gov/energyexplained/index.php?page=hydrogen_use

Experiment 23.4: Manufacture of Ethanol as a Fuel

A good experiment showing the production of ethanol by the fermentation of sugars by yeast. This has been a common practice for thousands of years and it is now being used in some sugar-producing countries, especially Brazil and Australia to manufacture ethanol as a fuel. In some countries, Australia included, it is sold mixed with petrol (10%). Overseas it is sold as gasohol.

The experiment requires a good concentration of sugar in solution, fresh yeast, a warm but not sunny position and time. Trials have suggested that at least 24 hours is required to give some alcohol but students will smell the sweet odour of the fermentation and see bubbles of carbon dioxide during the lesson. Hopefully one can obtain a professional chemical-laboratory glass distillation kit such as the 'Quickfit' system. Otherwise one will have to make up an older versions of fractional distillation apparatus with round flask, Liebig condenser etc,

QUESTIONS:

1. What is yeast?
Yeast are single-celled microorganisms that are classified, along with moulds and mushrooms, as members of the kingdom fungi.

2. Why was the mixture microwaved or mixed in warm water?
This was to ensure that the water is of the right temperature for the yeast to become activated. Personally, I would use warm water in case the microwaves kill the yeast.

3. What are some of the signs of fermentation occurring?
Bubbles of carbon dioxide will form and there will be a pleasant, sweet smell.

4. What was the purpose of the limewater? What did it show about fermentation?
It is used to detect the presence of limewater which will turn cloudy to show that carbon dioxide is given off during fermentation.

5. What was the purpose of the thermometer at the top of the apparatus?
When distilling the alcohol-water mix, the temperature should be kept about 80^{0C} as this is just above the boiling point of ethanol (78^{0C}) so that the alcohol and not the water is distilled off. This is why a water bath should be used in the heating by ensuring that the temperature does not go over 100^{0C}.

6. What is a condenser? Why was it important to have the water flow as indicated in the diagram?
The water-filled condenser is used to cool the alcohol vapour down to liquid form. The water is connected so that it runs up hill and not form a large air bubble at the top.

7. How was the condensate which came out of the distillation apparatus tested? What was the result?
Ethanol with minimal water and a drop placed on a watch glass will burn with a blue flame.

CONCLUSIONS:

1. What is fermentation?
Fermentation, is the chemical process by which molecules such as glucose are broken down anaerobically by microorganisms to produce carbon dioxide and alcohol.

2. Was any alcohol produced? Did it ignite?
This depends upon how well the experiment went! It should give a few drops on a small watch glass which can be ignited in a darkened room to see the faint blue flame. Some teachers (Heaven forbid! Have been known to add a little ethanol the night before!)

3. What was the original source of the alcohol?
The sugar and water which has been fermented.

4. Why was a water bath used and not direct heating with the burner?
So that the boiling temperature is kept below 100^{0C}.

5. What was the purpose of the thermometer at the top of the apparatus?
As before, this was to monitor the boiling of the mixture which should be kept about 80^{0C}.

6. What is a condenser? Why was it important to have the water flow as indicated in the diagram?
As before, this is to stop air bubbles at the top of the condenser.

RESEARCH: (Optional)

Use the internet to find out how ethanol is produced commercially and how it is currently used as a fuel. Mention the raw materials needed to produce the sugar and the advantages and disadvantages of using ethanol as a fuel

See:

https://afdc.energy.gov/fuels/ethanol_fuel_basics.html

https://www.nrcan.gc.ca/energy/alternative-fuels/biofuels/3493

http://large.stanford.edu/courses/2010/ph240/luk1/

http://biofuelsassociation.com.au/biofuels/ethanol/fuel-ethanol-blends/

https://www.fueleconomy.gov/feg/ethanol.shtml

23.4 Other Activities

- Produce methane gas by the decomposition of waste vegetable matter in warm water. In some rural communities, animal manure has been used in this experiment. It is smelly, messy but shows the principle of a biogas generator. Use the same set up as in fermentation of alcohol but collect the methane by displacement of water in the same way as one would collect hydrogen gas.

    http://www.saburchill.com/chemistry/chapters/chap047.html
    (use the over water method)

- Demonstrate a model fuel cell if one is available.

- Make a simple solar oven.

    https://sunshineonmyshoulder.com/6-homemade-solar-oven-projects-for-kids/

    https://www.instructables.com/id/Best-Solar-Oven/

    https://www.wikihow.com/Make-and-Use-a-Solar-Oven

- Make a simple solar hot water service using a hose coiled in a small, flat box and covered with clear plastic.

23.5 Answers to Multichoice Questions

Q1.A Q2.C Q3.A Q4.B Q5.A Q6.A Q7.D Q8.A Q9.C Q10.D

23.6 Some Suggestions for the Review Questions

1. **Considering the domestic use of electricity, what are the major appliances using power in the home? Are there any alternatives which can be used instead of electricity which would not create any other problems such as excessive gas emissions?**
   Any device which generates heat such as stoves, strip heaters, air conditioners and refrigerators use a lot of power. Evaporative cooling using a forced air draught (or even an open window) through an absorbent cloth is an alternative to air conditioning, albeit a poor one. Similarly, a Coolgardie Safe can be made as a box out of a light wooden frame covered with hessian bag with a tray of water on top and strips of wet material hanging down the sides. This is a simple cooler if placed in a shady spot with good air flow. It was used in outback homesteads to keep meat, milk and butter cool and worked quite well. Solar ovens may be able to do some limited cooking. See:

   https://lifeonspringcreek.com/2017/02/10/coolgardie-safe/

   https://permaculture.com.au/low-tech-refrigeration-solutions-the-coolgardie-safe-zeer-pot/

2. **Hydroelectricity is a major source of electricity in many countries e.g. Norway obtains about 95% of its power from this source. What could be some of the problems which such countries with such reliance on hydroelectricity face in the future?**
   The Norway reference may be inaccurate, but the principle is the same. They rely on good rainfall or snow melt and the construction of dams. If climate change produces less rain then the power stops.

3. **What are the major problems with the massive use of wind power? How can these be overcome?**
   Wind power may not be reliable with changes in wind systems if the strength of the wind is reduced. Regularity of electrical power can be maintained by storage batteries or producing hydrogen by electrolysis.

4. **The biggest problem with solar energy is that it only produces energy on clear, sunny days. Detail some of the ways by which solar energy could be stored for use at night.**
   The use of storage batteries, production of hydrogen gas or pumping water up into dams for later use as hydroelectricity.

5. **What is a fuel cell? How are fuel cells currently used?**
   Fuel cells are devices which generate electricity when fuels such as oxygen and hydrogen gases are added to a catalyst system. See:

   http://americanhistory.si.edu/fuelcells/basics.htm

https://www.hydrogenics.com/technology-resources/hydrogen-technology/fuel-cells/

6. **What are the advantages and disadvantages of planning to use coastal water power as a source of electricity?**
   Coastal power sources such as wave-action generators and tidal dams interrupt the usual natural and man-made activities of what usually is a very busy environment.

7. **Give several examples of how both plants and animals can be used in the production of energy.**
   Animals have long been a traditional source of energy for transport, turning grinding wheels and acting as power for lifting devices. Plant decomposition and as a source of alcohol by fermentation also have been used in the past.

8. **What are some of the current problems with using renewable energies to replace fossil fuels?**
   To date, they have not been able to come up to the amounts of energy needed. They are usually expensive in the initial phase.

9. **Consider an evening in a typical home during a winter season during the cooking of the evening meal. At that time, the following energy-using appliances are operating:**
   - stove with two hot plates on
   - solar water heater (no electrical booster)
   - a refrigerator (one-door)
   - an upright freezer
   - a log fire in a contained burner
   - a large television set (LED)
   - four LED lamps.

   If the evening meal preparation takes about one and a half hours, how much energy has been used in this time?
   A version of this exercise has been given beforehand in the Practical Manual.

10. **Do an energy audit of the power usage in (a) the home and (b) the workplace noting what appliances are used. Are there any ways in which this usage can be reduced using either no power or alternatives?**
    This is a useful. Personal exercise to be done personally or it could be done as a group activity.

23.7 Reading List

Aitken, D. W. (2010). *Transitioning to a Renewable Energy Future*, International Solar Energy Society,

Armaroli, N., Balzani, V. (2011). *Energy for a Sustainable World – From the Oil Age to a Sun-Powered Future*. London: Wiley-VCH 2011. ISBN: 978-3-527-32540-5.
Australian Government (2018). *Your Home – Australia's Guide to Environmentally Sustainable Homes*.
http://www.yourhome.gov.au/energy/renewable-energy
(Many good links)

International Energy Agency (IEA). 2014. FAQs: *Renewable Energy*.

Kaltschmitt, M., Streicher, W. & Wiese, A. (eds) (2007). *Renewable Energy. Technology, Economics and Environment*. Berlin/Heidelberg: Springer Berlin/Heidelberg. ISBN: 978-3-540-70947-3.

Luque, R., Lin, C., Wilson, K. & JClark, J. (Eds). (2016) *Handbook of Biofuels Production*. 2nd Edition. Amsterdam: Elsevier. ISBN: 9780081004555

Quaschning V. (2016): *Understanding Renewable Energy Systems*. 2nd edition. London: Earthscan. ISBN 978-113878-196-2.

Stolten, D. & Vikor Scherer, I. (Eds). (2013). Transition to Renewable Energy Systems. London: John Wiley & Sons. ISBN: 9783527332397

US Energy Information Administration (2018). Renewable Energy Explained.
https://www.eia.gov/energyexplained/?page=renewable_home
(Many good links about renewable energies)

Chapter 24: The Earth in Motion

24.1 Theme

The dynamic nature of Planet Earth is shown by slow and fast movements within its crust, atmosphere and hydrosphere. Volcanoes and earthquakes, landslides, avalanches, storms and wild seas and have been part of the life of many people in many countries around the globe. The Theory of Plate Tectonics has been a very useful model in explaining how the surface of the Earth is in motion.

24.2 Rationale

An understanding of the nature of forces within the Earth is extremely useful in engineering and risk management. Built structures such as dams, roads, bridges and buildings are subject to movement of the Earth. Often these forces and the movement that they cause are not obvious and sometimes they occur on a large scale with massive destruction to both the natural and man-made environment.

24.3 Notes on the Practical Work (see PRACTICAL MANUAL 2.)

Experiment 24.1: Hooke's Law

A simple experiment but it is an important one. Suitable springs are needed for the experiment. They must not be too stiff as to require very heavy weights nor must they be too light that they will stretch too quickly and be deformed. Springs and slotted masses and carriers are standard physics equipment but they can also be purchased at hardware stores. Spring Balances work in exactly the same way but using these removes the experimental aspect.

QUESTIONS:

1. Why is Hooke's Law valuable in understanding in how rocks fold or break along faults?
Within the elastic limit (i.e. when rocks will still rebound back to their original shape), stress (the forces applied) is proportional to strain (the deformation). It is important to understand how certain types of rocks will break or fold given specific amounts of forces.

2. What is the difference between ductile and brittle structure? Which apply to faults and folds?
Ductile structures such as folds show that they have bent due to external forces whereas brittle structures such as fractures and faults have been formed by the sudden breaking of rock.

CONCLUSIONS:

1. What is the shape of the graph?
It will be a straight line showing direct proportion.

2. What does this shape suggest about the relationship between the length of the spring and the force applied to the spring?
That they are in a direct relationship.

3. Is this consistent with Hooke's Law? Explain.
Yes, as it shows that the extension of the spring (its strain) is proportional to the force (stress) which causes it.

4. What is the value of the Spring Constant (k)?
This will depend on the type of spring used. This is found by measuring the slope of the graph. See:

http://www.4physics.com/phy_demo/HookesLaw/HookesLawLab.html

RESEARCH: (Optional)

Find out about the life and work of Robert Hooke. Why would knowledge of Hooke's Law be useful in geology?

Like many scientists of the day, Hooke was a polymath who was interested in many things. See:

http://www-history.mcs.st-and.ac.uk/Biographies/Hooke.html

http://www.ucmp.berkeley.edu/history/hooke.html

Experiment 24.2: Deformation of Rock and (the effects of) Temperature

Students like this experiment because they can use plasticine, play-dough or fresh putty (sufficiently wet) for this experiment. Again, some pre-experiment testing should be done to determine the water temperatures to use. See:

https://www.earthscienceeducation.com/taster/Deformation.pdf

QUESTIONS:

1. Why isn't plasticine considered elastic and obeying Hooke's Law?
Even when it is stressed (stretched) a little, it stays deformed.

2. Why was the plasticine placed into bags before emersion in water?
To keep it dry. Dry plasticine is better to handle even though it is water-proof.

3. Why is minimal handling required when attaching the plasticine to the stand and mass carrier?
The heat of the hand may make it softer.

CONCLUSIONS:

1. What is the relationship between temperature and the mass require to break the cylinder?
Increased temperature makes the plasticine more able to be deformed so it will stretch and break more easily. However, if plasticine is frozen it becomes more brittle and will break suddenly and cleanly. There is a good point here! Heated plasticine certainly deforms more easily and will continue to deform for a longer time when stressed but its strength is weaker at higher temperatures.

2. What is the relationship between the temperature and the time taken to break the cylinder?
It will take longer to break as the plasticine deforms for a longer period before breaking.

3. Describe how the cylinder broke for each of the temperatures.
There will be more deformation before the cylinder breaks.

RESEARCH: (Optional)

Other than external force and temperature, what other factors may be involved in the deformation of rock such as sedimentary rocks?
As well as confining pressure and temperature, the type of rock and time are the other main parameters of deformation. Sometimes sediments are pressured whilst they are still being consolidated and may contain water. These sediments then deform easily.

Experiment 24.3: Compressional Structures

This is best done as a demonstration as it requires some complex setting up. A small glass aquarium is ideal for this experiment and good, clean, well-sorted (sieved) sand is best. Once collected, the equipment can be cleaned and put aside for the next group.

QUESTIONS:

1. Why is it important to move the piston slowly?
These seems to be an overall elastic limit here as the sand/flour mix will suddenly break up if forced too quickly.

2. What is the purpose of the flour?
It acts as another type of sediment, gives contrast and may deform slightly differently to the sand.

CONCLUSIONS:

1. Comment on the use of this sandbox model in showing folding and faulting in rock.

This depends upon the performance of the experiment, but it generally is a good model to show deformation. The usual models for this type of activity are different layers of coloured foam rubber which can easily be deformed into the various types of folds but it is a simple demonstration only.

2. Relate the extent of the compression to the types of folding produced. Where in nature does this compression come from?

Given a long time, there is more deformation with extent of compression. The forces come through the Earth's plates due to plate movement, especially at margins of convergent boundaries.

3. How are such models useful in the understanding of a dynamic Earth?

Models allow researchers to apply ideas about large-scale observations by changing parameters and seeing if the results match reality. If so then the model becomes the theory for the observations.

RESEARCH: (Optional)

Use the Internet to find photographs of the types of folding and faulting seen in the model and then draw and label simple sketches of these.

See the pictures and diagrams in the textbook and the following:

http://www.geologyin.com/2015/02/types-of-folds-with-photos.html

http://earth.leeds.ac.uk/structure/learnstructure/index.htm

http://web.arc.losrios.edu/~borougt/GeologicStructuresDiagrams.htm

http://www.geo.hunter.cuny.edu/~fbuon/GEOL_231/Lectures/Fold-Fault%20Landforms.pdf

24.4 Other Activities

- Use coloured foam rubber strips to show folding

- Make paper models with drawn layers showing faulting and folding. Colour the different layers and given them the appropriate rock symbols. If a good size (say using large cardboard boxes) and covered in contact plastic, they can be used as a permanent display. See:

    https://www.fault-analysis-group.ucd.ie/papermodels/papermodels.htm

https://slideplayer.com/slide/4314540/

- Use layers of coloured plasticine or putty to make a layered sequence of sedimentary rock then push inwards from the side to show folding or cut with a knife and show faulting. The problem here is that one needs large amounts of plasticine to have a demonstration large enough or to have sufficient for students to make their own structures. Students also often like mixing the colours so recovery is often difficult. See:

  http://www.spacegrant.hawaii.edu/class_acts/FoldsFaults.html

24.5 Answers to Multichoice Questions

Q1.A Q2.D Q3.C Q4.A Q5.C Q6.B Q7.B Q8.A Q9.B Q10.C

24.6 Some Suggestions for the Review Questions

1. **List at least 5 pieces of evidence that the surface of the Earth is in motion.**
   Observed earthquake movement, slow deformation resulting in building movement, observed deformation as folds and faults, LASER geodometers across faults or spreading centres (Iceland), other plate movement detected by satellite GPS.

2. **What is a LASER geodometer? How is it used to measure movements in the Earth's crust? Give an example of such a measurement.**
   It is a very accurate distance-measuring device which fires a LASER beam across the suspected movement barrier to a reflector which sends the beam back to the geodometer. Knowing the speed of light, this measures the distance very accurately

3. **The following diagram is a view of a rock platform taken from above. It shows a particular type of joint pattern:**

What name is given to this type of pattern?
This is a conjugate joint pattern with joints generally crossing at about 60°.

Suggest the directions of the applied force causing the pattern.
The direction of the force which produces such patterns comes from the bisection of the angle between them. In this case the forces have come from a little off north and south as compression.

4. **What is Hooke's Law? How does it relate to the possible structures which may form in rock when pressure is applied? What factors could also be considered when thinking about which structures could form in the rock layers other than pressure?**
   Within the elastic limit, stress is proportional to strain. When rocks are subjected to applied pressure, slow application will produce folding. If the pressure exceeds the elastic limit of the rock it will be permanently deformed and may even suddenly break. Rapid application will usually break rocks as joints or faults.

5. **What are slickensides? What do they show about rock movement? What is the significance of serpentine?**
   Slickensides are seen as long parallel scratches on the fault plane. Often older faults will have mineralisation such as serpentine on the fault plane. This has come up from deep below, often as a hot solution. Serpentine will often have slickensides. The photo in the textbook (Figure 24.21) shows slickensides on serpentine in the valley of the Morado glacier taken by the author at about 3000 m in the Andes south west of Santiago, Chile.

6. **Use the textbook or Internet to locate the following major fault zones and describe their potential for future activity:**

   (a) **San Andreas Fault Zone** – San Francisco, USA. Great potential
   (b) **Alpine Fault Zone** – New Zealand. Moderate potential.
   (c) **Darling Fault Zone** – Western Australia west of Perth. Good.
   (d) **New Madrid Fault Zone** – south eastern USA. Good.
   (e) **Greendale Fault Zone** – South Island, New Zealand. Very good.
   (f) **North Anatolian Fault Zone** – northern Turkey. Very Good.
   (g) **Sunda subduction megathrust** – along the length of Sumatra, Indonesia. Great potential.

7. **In the field, how could one determine whether or not a fault was a tensional or compressional fault if the surface of the land had been eroded to a flat plain?**

   This is difficult if a side view has been obliterated by erosion. Having identified the strike of the fault, and knowing the sequence of rock layers, the geologist will have to identify the rocks on each side of the fault. Knowing the relative positions of layers on each side of the fault, a judgement can be made.

8. **What are transform faults? Why do they form and how important are they in the Theory of Plate Tectonics?**

   Transform faults are found at plate boundaries and are a succession of faults at right-angles to fracture zones. Movement along these faults is horizontal. They allow the movement of rigid, relatively flat tectonic plates to move over the curved surface of the Earth.

9. **What are some factors which can make the surface of a hill or mountain give way as a major landslide? In your explanation give some examples of landslides in various countries caused by one or more of these factors.**

   Landslides usually occur in nature when the soil or rock becomes too heavy for their slope to remain stable. Landslides can also be initiated by earth movements such as small tremors to large earthquakes. Landslides can also occur if man-made buildings are added to unstable slopes, especially if there is water flowing through the soil. Thixotropic clays also give way suddenly with slight vibrations. Examples are many but the best known in Australia is the Thredbo disaster of 1997 when a landslide collapsed several ski lodges in the Thredbo resort killing a number of people. It was thought that poor water drainage assisted in the land giving way under the buildings.

10. **In some places in the European Alps, 19th century geologists found a sequence of rock strata which seemed to have the oldest rocks on top of youngest rocks. This was against the well-established Law of Superposition. Explain how this sequence could have occurred.**

    This confounded early geologists who were well trained in the Law of Superposition that oldest beds were on the bottom of a sequence and youngest beds on top. In the European Alps, some rock layers showed ages which went against this law. Further, more extensive mapping suggested that the geologists had been looking at only a small section of an overfold which had been eroded so that the full folding was not obvious.

24.7 Reading List

Burbank, D.W. and Anderson, R.S. (2011): *Tectonic Geomorphology*, 2nd Edition. Wiley-Blackwell. 427 pp. ISBN: 978-1-4443-3887-4

Davis, G. H. and Reynolds, S. J. (1996). *Structural Geology of Rocks and Regions.* New York, John Wiley & Sons. pp. 372-424. ISBN 0-471-52621-5.

Hancock, Paul L., Skinner, Brian J., Dineley, David L. (Edits.) (2000). *The Oxford Companion to The Earth.* 1184 pp. Oxford University Press. ISBN 0-19-854039-6.

Park, R.G. (2004). *Foundation of Structural Geology (3 ed.),* Routledge. 214 pp. ISBN 978-0-7487-5802-9

Pollard, D.D. and Fletcher, R.C: (2005). *Fundamentals of Structural Geology.* Cambridge University Press. ISBN 0-521-83927-0.

Skinner, B.J., Porter, S.C. & Park, J. (2004). *Dynamic Earth - An Introduction to Physical Geology* 5th edition. 236 pp. John Wiley & Sons, Inc.

Also see:

https://pangea.stanford.edu/projects/structural_geology/

http://www.geologyfieldtrips.com/hayward.htm

https://earthquake.usgs.gov/learn/topics/shakingsimulations/

Chapter 25: Volcanoes

25.1 Theme

Volcanoes show that planet Earth is an active planet with a changing surface. Volcanoes are sources of massive destruction as well as rich soils and abundance. Different types of plate activities cause different types of volcanoes.

25.2 Rationale

Many countries have active volcanoes and others show the results of ancient extinct volcanism. New volcanoes are also being produced around plate margins and at certain hot spots within plates. An understanding of how volcanoes work assists in learning about the dynamic surface of the Earth. Many people live and work on nearby active volcanoes and there are many career pathways in the science of volcanology and the technologies used to monitor and predict volcanic eruptions.

25.3 Notes on the Practical Work (see PRACTICAL MANUAL 2.)

Experiment 25.1: The Shape of a Volcano

This is a good exercise in revising the drawing of topographical cross-sections and then interpreting the shape of the landforms. This map is actually a modified plan view of Mount Pinatubo in the Philippines. It is an active stratovolcano being constructed of both ash and lava. It had a massive Plinian eruption (i.e. mostly ash and surface clouds of gas and ash as pyroclastic flows) in 1991. Whilst its slopes have been greatly eroded by tropical rains, it still shows the relatively uniform slopes and cone shape of such volcanoes with some steepness due to the more recent ash falls.

The final shape of the volcano will look something like:

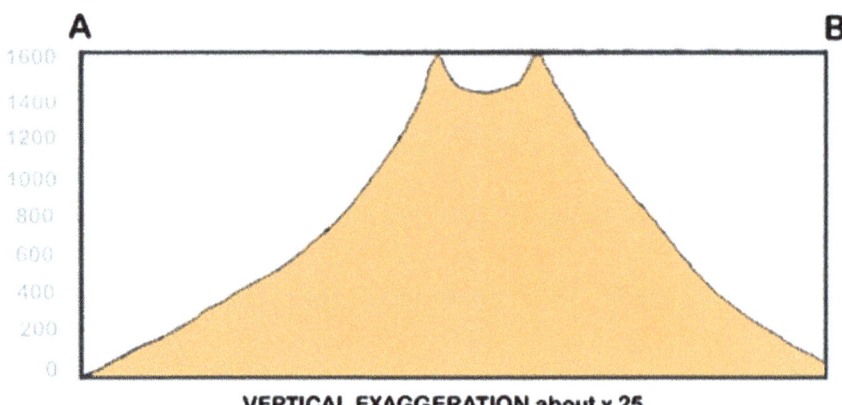

VERTICAL EXAGGERATION about x 25

See:

https://www.ngdc.noaa.gov/hazard/stratoguide/pinfact.html

http://mountpinatubo.net/

https://pubs.usgs.gov/fs/1997/fs113-97/

QUESTIONS:

1. What are the horizontal scale and vertical scales?
Depends upon how the students draw the cross-section. The horizontal scale is given on the map and a recommended vertical scale is suggested at 1 cm = 200 metres of height.

2. What is the approximate Vertical Exaggeration in numbers and what is its meaning? Why is this important?
This depends upon the cross-section drawn by the students. On the constructed section above, it is about 25.

See:

http://community.boredofstudies.org/23/geography/241383/how-calculate-verticalexaggeration.html

CONCLUSIONS:

1. How is this volcano most likely to be classified (as far as eruption material is concerned)?
Stratovolcano formed by alternative layers of ash and lava (andesite and dacite for Mt. Pinatubo)

2. Give reasons for your classification.
The uniformity of the gradient of the slopes, especially at lower heights and closer to B.

3. If the volcano has not erupted in many years, what would any new eruption likely to be in the first few hours?
These volcanoes (including the prototype Mt Vesuvius) usually explode with little warning giving a typical Plinian eruption of large volumes of gas, steam and ash being erupted to great heights.

See: http://www.geology.sdsu.edu/how_volcanoes_work/Plinian.html

4. What would be the main hazards to the nearby cities? Why?
The main hazards are from lahars (volcanic mudflows) and ash. For Mt. Pinatubo, most of the surrounding plains consist of deep lahar deposits. Lahars flow very quickly down the gullies eroded into the slopes of the volcanoes and would quickly engulf many of the towns out on the plains (e.g. Angeles City). Ash falls would also

be extensive downwind of the volcano (probably towards the southwest – north at the top of the map – towards San Mercelino).

RESEARCH: (Optional)

1. Use the Internet to find out how volcanoes (especially those near centres of population) are monitored and predictions made about future eruptions.
Direct observation, tiltmeters to measure changes in slope, seismographs to measure magma movement, gas monitors to detect increased gas emissions, monitoring the depth of the water table which changes with ground movements and some people claim that animal activity can also predict volcanic events. See:

https://volcanoes.usgs.gov/vhp/monitoring.html

https://www.bbc.com/bitesize/guides/z8p9j6f/revision/5

2. What emergency action would be taken by the local population if there was an imminent threat of this volcano erupting?
See the emergency plan in the textbook. Even students who do not live near an active volcano should be aware of what do in a sudden eruption, especially with Asian and South American volcanoes.

Experiment 25.2: Locations of Some Major Volcanoes

Another good latitude-longitude plotting exercise which should be related to the later mapping of major earthquakes so that students should see that volcanoes and earthquakes are found in the active zones along the tectonic plate margins.

PART A: Locations of some active volcanoes

QUESTIONS:

1. How could this plot be improved?
Use of more data with many more active volcanoes (over 1500 in reality) and use a bigger map. Teachers might like to reproduce a world map with coordinates on an A3-sized paper.

2. What are the criteria which determine the activity state of a volcano?
Volcanoes are considered active if there has been an eruption in historical times. It is dormant if it has been known to have been active but has not erupted for some time (a few years). It is extinct if there has been no historical record of any eruptions, especially if it has been eroded.

PART B: Volcanoes and climate change

QUESTIONS:

1. What is the general trend of this graph?
It is a direct proportion graph showing that global temperature is rising.

2. What does this graph generally show about climate change?
With some sudden cool changes excepted, the climate is warming.

CONCLUSIONS:

1. Is there any pattern or logical groupings to the locations of the volcanoes?
With some exceptions of mid-plate volcanics, students should be able to group those volcanoes of the Pacific ring of fire.

2. Why do these volcanoes exist in these locations?
Volcanoes occur at or near the edges of the tectonic plates with andesitic stratovolcanoes at subduction zones, basaltic shield volcanoes within plates and at mid-ocean ridges and some rhyolitic types inland on continents also due to subduction.

3. Account for the volcanoes in other locations where there might not be a grouping (e.g. Hawaii)?
These are mostly due to re-melting of crustal basalt due to increased heat at hot spots above mantle plumes. Lines of volcanoes show the movement of the plate over these plumes.

4. How could a more accurate world pattern be obtained, especially in locations of isolated volcanoes?
More detailed and accurate mapping with more locations is required.

5. Is there any relationship between some volcanic eruptions and climate change? Explain.
The sudden down turn in the graph represents major volcanic eruptions which have put so much ash and other particulate matter into the atmosphere that they have caused wide-spread cooling as the sunlight was blocked.

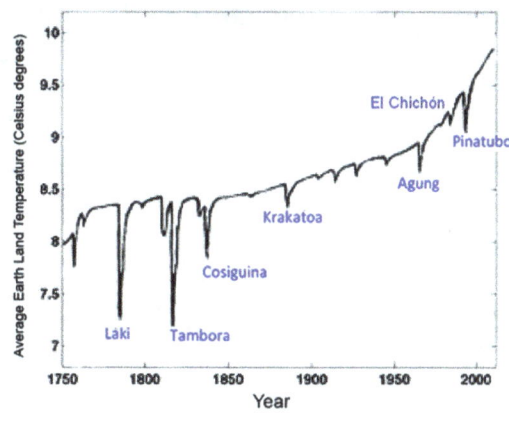

6. Could such a change occur in the future?
Unpredictably yes, from one of the many super volcanoes around the globe, especially those in Iceland, Asia and others.

RESEARCH: (Optional)

Use the Internet to:

1. locate currently erupting volcanoes.

2. locate any places where a major volcano may erupt e.g. a region with a large amount of heat which could cause the eruption of a new or existing dormant volcano.

See:

https://www.volcanodiscovery.com/erupting_volcanoes.html

https://www.usgs.gov/faqs/how-many-active-volcanoes-are-there-earth?qt-news_science_products=0#qt-news_science_products

https://www.dkfindout.com/us/earth/volcanoes/where-are-earths-volcanoes/

25.4 Other Activities

- Model volcanoes can be a bit messy and students may lose the connection with the real thing, but they are motivating. The simplest type is using bicarb soda (sodium bicarbonate) and vinegar to make a messy foam which will flow over a wide area simulating lava. Add red food dye to the bicarb and place it in a small container in the "crater" of a mound of sand or modelling clay (to represent the volcano). This should be done in a wide tray or outside depending upon the size of the model. There are other versions, but they usually involve the emission of toxic fumes and carcinogenic ash!

- Another model to simulate extensive ash fall uses ammonium dichromate (with a little added sulfur and iron filings to give a realistic smell and sparks). A small pile of this chemical (see the Chemistry Faculty) is placed on top or in a shallow hollow of the clay or sand volcano. When ignited with a match, a very large volume of green chromium III oxide is given off as a fine ash (and some sulfur dioxide if sulfur is used). CARE: the ammonium dichromate is carcinogenic and the ash and fumes toxic. This is best done in a large fume hood or outside – it will be very messy but is a good simulation of ash fall. If considered too dangerous, watch the video:

    https://www.youtube.com/watch?v=Ula2NWi3Q34

    Also see:

http://www.rsc.org/learn-chemistry/resource/res00001709/ammonium-dichromate-volcano?cmpid=CMP00005223

https://www.angelo.edu/faculty/kboudrea/demos/volcano/volcano.htm

- If available, examine some recent volcanic ash under a stereomicroscope and see its sharp-edged shape. Also feel the gritty nature of this fine grained material. Discuss its effects on breathing and getting into aircraft engines at height and also into other machinery.

- Watch the video on Pompeii from the textbook. Discuss the effects of an ash cloud rolling into a large city (note: modern Pompeii is a suburb of Naples, Italy – it has narrow streets, is overcrowded and not far from Mt. Vesuvius). See:

Online Video: Explore the ruined city of Pompeii, Naples, Italy
Go to https://youtu.be/PevacwLMvWU

- Also watch the eruption of Yasur on Tana Island, Vanuatu in the Pacific witnessed by the author. See:

Online Video: Travel to Yasur, the active volcano of Vanuatu in the south Pacific and drive over the moon-like landscape
Go to https://youtu.be/oML8kdnfLYU

- Have the students look for the locations of volcanoes in out-of-the-way places such as Antarctica etc. See the video from the textbook on Deception Island taken by the author in 2011:

Online Video: Sail into the caldera of Deception Island, Antarctica - a dormant stratovolcano
Go to https://youtu.be/Y0z2MLQ391U

- Discuss secondary volcanic features such lava tubes, geysers, solfatara, hot springs and mud pools. See the remaining videos taken by the author in Chapter 15 of the textbook.

- Watch the movie Dante's Peak – good Hollywood fiction but many of the precursors, equipment and volcanic events are based on real events (Dante – the robot, not the author – was used in a crater in Antarctica but it too failed). Students can discuss the fiction and non-fiction of the movie.

25.5 Answers to Multichoice Questions

Q1.D Q2.B Q3.C Q4.C Q5.D Q6.D Q7.D Q8.A Q9.A Q10.C

25.6 Some Suggestions for the Review Questions

1. Distinguish between each of the following:

    (a) **extinct and dormant** - extinct will never erupt again but dormant may.
    (b) **crater and caldera** - crater is the circular depression of a volcano but caldera is a much larger depression caused when the magma chamber below collapses.
    (c) **lava and magma** - lava is molten rock flowing on the surface and magma is the pool of molten rock deep below the volcano.
    (d) **extrusive igneous and intrusive igneous** - extrusive igneous rocks form on the surface and intrusive rocks cool below the surface.
    (e) **basic, acidic and andesitic magmas** - acid igneous rocks have high silica content whereas basic rocks have little or no silica. Andesitic rocks have a moderate amount of silica and andesine plagioclase.

2. Define the meaning of each of the following:

    (a) **lahar** - is a fast-flowing mud stream formed when volcanic ash is mixed with water
    (b) **fumarole** - is a volcanic gas vent
    (c) **volcanic bomb** - is an ejected wad of lava which forms into a twisted shape as it moves through the air.
    (d) **parasitic cone** - is a small cinder vent which is forces out onto the slopes of a volcano.
    (e) **solfatara** - fumarole giving off sulfur vapour.

3. Use the internet and determine the specific danger or problem for animal life from:

    (a) carbon dioxide
    (b) volcanic steam
    (c) sulfur dioxide
    (d) volcanic ash

    All of these will cause suffocation in animals and death to plants in larger amounts. Some ashes are also toxic and will cover plants and kill animals when the plants are eaten. Carbon dioxide is particularly dangerous as it has no odour. Over-turning of decayed plant matter in crater lakes have been known to give off much carbon dioxide which flows downhill killing all animal life.

4. What is a nuée ardente? What are its characteristics? Give examples of eruptions where these have been a common form of destruction.

From the French for glowing cloud, these are a highly destructive, fast-moving, incandescent mass of gas-enveloped particles that is associated with certain types of volcanic eruptions. The term was coined by a French geologist observing the eruption of Mount Pelée on the island of Martinique in 1902 but they common elsewhere such as in the Philippines at Mount Pinatubo and Mount Mayon, in Japan at Mount Unzen and have also occurred in the past at the eruptions of Mount Vesuvius and Krakatoa. They are now called pyroclastic flows and consist of a hot (600^0 +) foam of ash, steam and other gases which can be ejected from the tops or side vents of volcanoes of speeds of several hundred kilometres per hour.

5. **What is a Plinian eruption? Which is the most destructive phase of such an eruption? Use the Internet to find out about Pliny the Elder and the destruction of Pompeii (Pliny the Younger, his nephew gave an excellent description of Vesuvius and his uncle's activities.**
It is a very destructive eruption with a large volume of ash being driven to great altitudes but then collapses. This is usually the first, violent eruption of a dormant stratovolcano. It begins with a sudden, rapid ejection of a huge amount of ash, pumice, larger particles and gas straight upwards to great heights (sometimes many tens of kilometres into the atmosphere). This is usually extremely loud and there is a considerable amount of lightning within the column due to static electricity caused by friction. The column may collapse giving large pyroclastic flows which roll down the sides of the volcano and travel out (even across water) to great distances and thickness. There may then be a period of smaller pyroclastic flows coming from the vent or side eruptions of the volcano. Then there may be a relatively quiet period followed by lava flows.

The eye-witness account by Pliny the Younger in 79 AD was a testament for his uncle, Pliny the Elder, who was the Roman Admiral who tried to save the peoples of Pompeii and Herculaneum but died in the attempt. See:

http://volcanology.geol.ucsb.edu/pliny.htm

http://www.eyewitnesstohistory.com/pompeii.htm)

6. **The stories of the birth of volcanoes are often good examples of how fragile the Earth's crust and how suddenly conditions upon it change. Research the sudden birth of the volcanoes:**

   (a) **Paricutin in Mexico** – a large cinder cone which surged suddenly from the cornfield of local farmer in 1943.

   (b) **Surtsey in Iceland** – this new island formed from a volcanic eruption which began 130 metres below sea level just south of Iceland and reached the surface on 14 November 1963

   (c) **Hunga Tonga volcano in the Pacific** – this formed suddenly from an older submarine volcano forming a new ash island, unofficially named Hunga Tonga-Hunga Ha'apai. The new island formed during a submarine volcanic eruption that lasted from late December 2014 to early January 2015. The new land mass, which has a 120m summit, was originally only predicted to last months but still exists.

7. **Iceland has been called the land of fire and ice. What is unique about several of the volcanoes in Iceland? Why are they so dangerous?**

   Icelandic volcanoes form at the mid-ocean ridge spreading centre which cuts through the island. These are mainly a basaltic composition, erupting as long fissure volcanoes with many high fire fountains of very hot (1000^0C +) lava. Some of them form below thick icecaps and glaciers with the result that huge volumes of water suddenly form and flooding of parts of the island is common. Some eruptions also eject large amounts of steam and ash into the upper atmosphere which can carry for thousands of kilometres, disrupting air traffic.

8. **Antarctica is a continent which is almost covered in ice but it too has volcanoes. Use the internet to locate and describe the largest active volcanoes of Antarctica.**

   A study in 2017 suggested that there could be up to 138 volcanoes below the thick Antarctic ice sheet. Deception Island near the Antarctic Peninsula is a ring-shaped caldera which erupted in 1965 and 1967 (see the author's video given above) and Mounts Erebus and Terror form most of Ross Island in western Antarctica. See:

   https://www.auroraexpeditions.com.au/blog/hot-and-cold-volcanoes-in-antarctica

   https://www.volcanodiscovery.com/antarctica.html

9. **Volcanic eruptions can be difficult to predict. What are some warning signs and measurements which could be taken to suggest that an eruption is imminent?**

   These can include small harmonic seismic tremors (regular vibrations), changes in the slope of the land, changes in water levels, emission of radon gas, increased temperatures of lakes and ponds and sometimes unusual behaviour of animals who flee the area.

10. **Use the Internet to assess the volcanic hazard potential of each of the following places:**

    (a) **Mexico City** - Popocatépetl is 70 km southeast of Mexico City,
    (b) **Naples, Italy** – Mt. Vesuvius is on the outskirts.
    (c) **Kalapana, Hawaii** – suffered recent lava coverage from Kilauea
    (d) **Kagoshima, Japan** – Mt Sakurajima is nearby.
    (e) **The local area** - ? If there are no active volcanoes, research the nearest <u>extinct</u> volcano.

    Places (a) to (d) have had past volcanic activity and exist in active volcanic regions. See previous maps of world's active volcanoes.

25.7 Reading List

Decker, R. W. & Decker, B. (1991). *Mountains of Fire: The Nature of Volcanoes*. Cambridge University Press. p. 7. ISBN: 0-521-31290-6.

Hamilton, J. (2010). *Volcano*. Compton Verney, UK. ISBN-10: 0955271959

Hull, E. (2015). *Volcanoes Past and Present*. Colorado Springs., CO. CreateSpace Independent Publishing Platform. 142 pages. ISBN-10: 1505584779

Marti, J.& Ernst, G. (2005). *Volcanoes and the Environment*. Cambridge University Press. ISBN: 0-521-59254-2.

Scrope, G.P. (2009). *Volcanoes.: The Character of Their Phenomena, Their Share in the Structure and Composition of the Surface of the Globe, and Their Relation to Its Internal Forces*. University of Michigan Library. 522 pages. ASIN: B002KT3FCY

Chapter 26: Earthquakes

26.1 Theme

Earthquakes occur when there is a sudden breakage of rock or movement along a fault plane or plate margin due to the rock being stressed and deformed then suddenly releasing energy. This is the elastic rebound theory. They can be slight tremors, or major events causing large-scale damage by collapse of buildings, roads and dams, landslides and ground upheaval. Tsunamis formed by vertical movement under the sea also cause massive damage and loss of life along coastal regions.

26.2 Rationale

To understand how earthquakes occur is to understand the sub-surface nature of the Earth's crust and deeper interior. Earthquake prediction is still not reliable but an important part of many countries and their economies.

26.3 Notes on the Practical Work (see PRACTICAL MANUAL 2.)

Experiment 26.1: Earthquake Epicentres and Magnitudes

This is a very thorough experiment, simulating what would go on if seismograms where used manually. Modern seismic recording centres will do all of this calculation by computer and the epicentre and its magnitude will be known instantly. The movement of the Earth varies across the globe and earthquakes are a regular, daily event in many places but these usually go unfelt.

PART A: Locating the epicentre.

QUESTIONS:

1. How are the P and S waves detected at each of the stations?
By seismographs (seismometers) which produce seismograms on computer screens. Once these were recorded on revolving drums called kymographs with ink pens tracing the seismogram on paper. Even earlier seismographs had scrapers, not pens which scraped the seismogram on a carbon-coated drum.

2. Why does the P wave arrive before the S wave?
Because it travels much faster than the S wave but the actual speeds depend on the rocks through which they both travel. See:

http://eqseis.geosc.psu.edu/~cammon/HTML/Classes/IntroQuakes/Notes/waves_and_interior.html

3. Comment on the need for accuracy in making measurements from seismograms.

Using this paper method, the scale would mean that a slight misjudgement of the travel-time distance will give a poor measurement of distance. This is the main error in this experiment which will be shown in the size of the triangle of error where the circles around the seismic stations are drawn.

4. How could this accuracy be improved?

Closer measurement of the P-S travel-time distance on the seismograms. If they can be enlarged electronically or expanded using a copier then this would help.

PART B: Finding the magnitude of the earthquake.

QUESTIONS:

1. In general, the amplitude should get smaller with distance from the epicentre. Why?

As any wave travels outward from its source, it loses energy which is shown by a decrease in its amplitude.

2. What other factors other than distance could reduce the amplitude of an earthquake wave?

The nature of the rock and also how much water would be contained in any porous rock.

3. Would such as earthquake trigger a tsunami? Why? Not all earthquakes near the sea produce tsunamis. Why?

There is no indication that this earthquake would cause a tsunami as the experimental data would not show this. Other data in real seismology would show the vertical component as well. This experiment is a theoretical construction based on the real earthquake and tsunami which devastated the Palu region on September 28, 2018. See:

https://phys.org/news/2018-10-indonesia-tsunami-worsened-palu-bay.html

https://theconversation.com/reviewing-indonesias-tsunami-early-warning-strategy-reflections-from-sulawesi-island-104257

Not all earthquakes, including those from foci under oceanic crust will cause a tsunami. Tsunamis are caused by vertical displacement and some of the worst tsunamis came from landslides and not earthquakes, although the two may come at the same time e.g. the tsunami in Indonesia in late 2018 due to the collapse of part of the erupting volcano Anak Krakatoa. In 1958, an earthquake in Alaska triggered a massive landslide at the end of the long inlet of Lituya Bay which produced a megatsunami which reached a height in the confined bay of 520 metres.

CONCLUSIONS:

Give a report on the location and magnitude of this earthquake in Indonesia. Also suggest other problems which may occur as a result of this earthquake near the sea.

The epicentre of the earthquake in this exercise is centred about 100 km north of Palu, Indonesia and just off the coast. Its magnitude is about 7.0 on the Richter Scale.

RESEARCH: (Optional)

Find out if an earthquake did occur in this region about September, 2018 and describe some of the problems associated with the earthquake as well as the amount of destruction, casualties and any actions which may have reduced the damage. Give a Modified Mercalli Scale value to this earthquake. Why was this earthquake especially bad?

The real earthquake and its tsunami in the Palu region (see links above) was rated at 7.5 Richter Scale and about 6 -7 on the Modified Mercalli Scale, although this will vary from place to place. See:

https://www.iris.edu/hq/files/programs/education_and_outreach/retm/tm_180928_indonesia/180928Indonesia.pdf

Experiment 26.2: Locations of Some Major Earthquakes

This is a useful paper plotting exercise which gives good revision in the plotting of latitude and longitude. The exercise may become boring in its initial stages so students should be encouraged to look for patterns on the map. These will have to be photocopies for the students and those students who wish to submit electronic reports can scan or photograph their finished maps.

QUESTIONS:

1. How could a more accurate world pattern be obtained, especially in locations of isolated earthquakes?
Use much more location data, more accurate latitude and longitude values and a bigger map.

2. Account for the epicentres in other locations which do not appear to have major earthquakes according to this data.

These are intra-plate earthquakes which are often caused by sudden movement along older fault lines within the plate. These are triggered off by pressure which has been applied through the plate by its movement. Earthquakes in Australia and the interior of the United States are of this type.

CONCLUSIONS:

1. Is there any pattern or logical groupings of epicentres?
The final plot should show lines (or curves) of epicentres around the tectonic plate margins.

2. Why do these epicentres exist in these locations?
These epicentres correspond to major subduction zones or conservative boundaries.

3. How do these earthquake locations compare with those of volcanoes from Experiment 25.2?
The locations should show a great similarity as the andesitic type of volcano is also formed on the downward side of subduction zones.

4. From the map and other sources, list a few major cities (in) which there could be valid predictions for:

       a. earthquakes
       b. tsunamis and
       c. volcanoes

For sections (a) and (b) these cities could include any around the Pacific ring of fire, including San Francisco, Anchorage, Los Angeles, Lima, Tokyo, Djakarta and even Perth and Broom in Western Australia.

Cities which may have volcanic eruptions nearby include any of those in Japan, the Philippines, along the coastline of the Andes and north western United States and Alaska. The destruction of part of south eastern Hawai'i has shown the effects of Kilauea, a mid-plate volcano.

5. Comment on each of the following statements:

    a. **Earthquakes are caused by volcanoes** – whilst magma movement below volcanoes can cause tremors, large earthquakes are caused by plate movement. They often are in the same area because some volcanoes and earthquakes are caused near subduction zones of plates.

    b. **Volcanoes are caused by earthquakes** – Generally eruptions are caused by other events but some scientists think that in some cases this may happen as earth movements may disrupt the magma chamber below. See:

11. https://www.usgs.gov/faqs/can-earthquakes-trigger-volcanic-eruptions?qt-news_science_products=0#qt-news_science_products

b. **Earthquakes and volcanoes can have common sources** – in many cases this is true. Volcanoes and earthquakes are associated with movement of the Earth's plates, especially at subduction zones.

RESEARCH: (Optional)

Use the Internet to find out about some other earthquakes e.g.

- Lisbon, Portugal in 1755
- New Madrid, USA in 1812
- San Francisco, USA in 1906
- Southern Chile in 1960
- Meckering, Australia in 1968
- Tanshan, China in 1976
- Loma Prieta , USA in 1989
- Newcastle, Australia in 1989
- Kobe, Japan in 1995
- Gujurat, India in 2001
- Haiti 2010
- Pueblo, Mexico in 2017

See:

http://oxfordre.com/naturalhazardscience/view/10.1093/acrefore/9780199389407.001.0001/acrefore-9780199389407-e-303

https://en.wikipedia.org/wiki/Lists_of_earthquakes

https://ourplnt.com/top-10-most-powerful-earthquakes-in-recorded-history/

https://earthquake.usgs.gov/earthquakes/browse/largest-world.php

https://moneyinc.com/20-costliest-earthquakes-world-history/

Also see:

https://ourworldindata.org/natural-disasters

https://www.shakeout.org/

http://quake.utah.edu/regional-info/earthquake-faq

Current earthquake monitor sites:

http://ds.iris.edu/seismon/?

https://www.emsc-csem.org/#2

https://earthquakes.ga.gov.au/
(Australia and the world)

Experiment 26.3: Predicting Earthquakes

This is a good exercise in understanding how some of the ordinary work of scientists can add together to give a larger picture in an attempt to predict earthquakes. One of the main ways in which seismologists attempt to make predictions is by measuring the regular stress and movements across active fault line.

QUESTIONS:

1. What is a LASER geodometer? How does it work?
This is a very accurate distance-measuring device which fires a LASER light beam to a target reflector at some distance. The light reflects to the main geodometer and the distance to the target can then be calculated.

2. What is a cumulative right-lateral creep?
This is the continually-added distance moved along the faultline along its strike i.e. sideways or laterally along the surface. Faults which move horizontally are known as strike-slip faults and are classified as either right-lateral or left-lateral. In a horizontal point of view, a right-lateral movement is seen if one were to stand on the fault and look along its length, then the right block moves toward you and the left block moves away.

See:
https://earthquake.usgs.gov/learn/glossary/?term=right-lateral

https://www.usgs.gov/faqs/what-a-fault-and-what-are-different-types?qt-news_science_products=0#qt-news_science_products

3. How will such measurements help with probabilities of future earthquakes?
The movement is also a measure of the stress applied at various points along the length of the fault. Measurements which suggest that stress is building up at one particular point along a faultline can infer that this may be released soon and cause an earthquake. Along a particular faultline, seismologists might have some idea of the maximum stress which will occur before the fault suddenly moves.

CONCLUSIONS:

1. **What does the graph suggest about the possibility of an earthquake along the Calaveras Fault?**
   The graph should look something like:

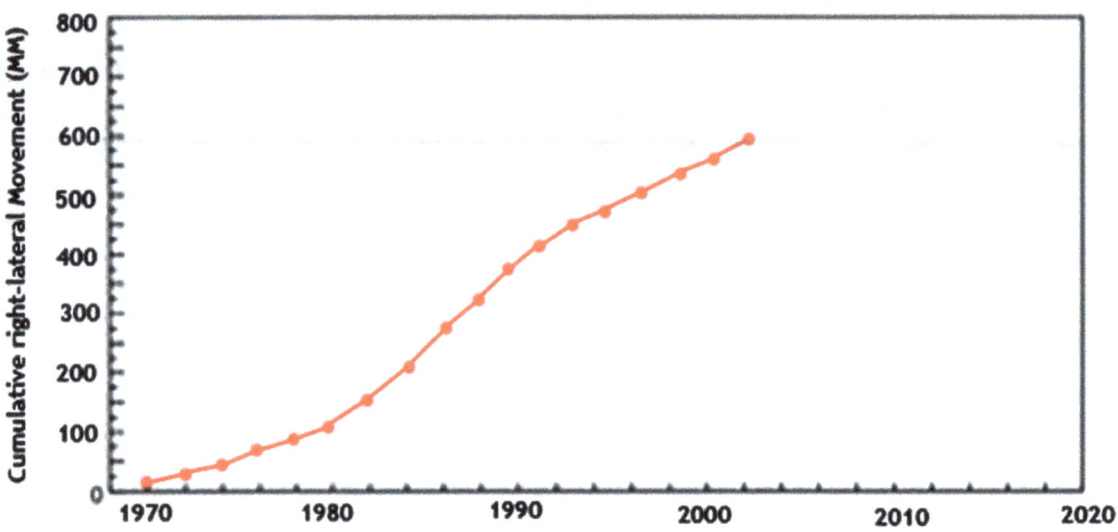

This shows a general increase in movement (and stress) over time with a more dramatic movement after 1982 suggesting a probably of potential for an earthquake.

2. **Comment on the use of such measurements in assisting with earthquake prediction. Comment on any correlation with Californian earthquakes since 1970 and their relationship to the graph.**

   Measurements from such graphs may give an indication of potential earthquakes when there is a sudden change in the gradient of the graph indicating that stress is causing a greater-than-average movement. The Morgan Hill event occurred in 1984. April 24 at 1:15 p.m. local time in the Santa Clara Valley of Northern California. The shock had a moment magnitude of 6.2 and a maximum Mercalli intensity of VIII (*severe*). The epicentre was located near Mount Hamilton in the Diablo Range of the California Coast Ranges. Nearby communities (including Morgan Hill) sustained serious damage with financial losses.

3. **Also assess the difficulty which geophysicists would have in attempting to measure data in the attempt to predict earthquakes.**

   With such measurements as those given in this exercise, seismologists can only infer that a slight increase in movement may suggest that something is happening. It is always possible in hindsight to say that such a change in the graph will predict an earthquake, but this is not always the case before the event. Each faultline will behave differently. In general, there are many other factors which make earthquake prediction difficult.

RESEARCH: (Optional)

Use the Internet and reliable scientific sites to make an evaluation of the ability to predict earthquakes, volcanic eruptions and tsunamis.

See:

https://www.livescience.com/56487-why-earthquake-prediction-remains-elusive.html

https://www.geosociety.org/awards/14speeches/GML-PresMedal.pdf

https://www.usgs.gov/faqs/can-you-predict-earthquakes?qt-news_science_products=0#qt-news_science_products

https://theconversation.com/why-it-is-so-hard-to-predict-where-and-when-earthquakes-will-strike-40873

26.4 Other Activities

- Make an electronic seismograph using a strong bar magnet suspended by a loose spring. The magnet should be able to fit into a hollow, insulated coil, the end of which should be connected to a Cathode Ray Oscilloscope. When the CRO is adjusted for maximum sensitivity for amplitude, a slight up or down motion of the magnet (due to vibration of its stand) will show a good amplitude. If the horizontal sweep of the CRO is also set so that the trace moves across the screen, a good representation of an earthquake wave is seen. The equipment (and a helper) may have to be borrowed from the Physics faculty.

- View the video of a seismogram from the textbook at:

     Online Video: An example if an incoming seismogram
    Go to https://youtu.be/wp3H6GZ_TdE

- Demonstrate waves through slinky springs to show longitudinal and transverse wave differences. Students should note that it is energy travelling through the spring, not the material of the spring itself.

- Build an earthquake-proof structure and see if it withstands a shake! See:

    http://teachers.egfi-k12.org/activity-earthquake-proof-structure/

- Do an audit of the classroom and school building construction and discuss its potential for collapse in major earthquake (e.g. brick buildings fall apart, concrete is more stable, glass windows pop out etc.). Does the institution have an earthquake emergency procedure? Use the Internet to find out the potential of the local area for major earthquakes.

26.5 Answers to Multichoice Questions

Q1.A   Q2.C   Q3.D   Q4.B   Q5.B   Q6.C   Q7.C   Q8.B   Q9.C   Q10.C

26.6 Some Suggestions for the Review Questions

1. **Earthquakes are very difficult to predict. What are some indications (scientific and others) that may indicate that an earthquake might occur soon in a local area? Discuss.**
   Often controversial and more research needed, but some precursors of earthquakes mentioned in the literature include: increased seismic tremors on local seismographs; increases in stress as measured by strain devices over faults; unusual activity of animals (especially those caged); emission of radon gas (radioactive); some changes in the local electromagnetic field; fluctuations in ground water levels in wells etc.

2. **What is a seiche? Could it be a major earthquake hazard? Explain.**
   Seiches are water waves in confined waters such as pools and lakes which vibrate back and forth as standing waves due to the water's resonance with earthquake waves. They can also be caused by a consistent wind across a lake. They can be hazardous if the waves are high and can cause problems with water navigation and local constructions such as bridges and dams.

3. **Distinguish between P, S and L waves. Which of these waves would not travel through water? What is the significance of that?**
   P (Primary) and S (Secondary) waves are body waves which travel through the Earth. L (Love - named after A.E.H. Love) waves travel around the surface and are most destructive. P waves are compressional waves and can move through solids and fluids. S waves are transverse waves which can only pass through solids. Surface waves include L waves which have a horizontal motion and Rayleigh Waves which have a rolling motion.

4. **Discuss the terms magnitude and intensity with reference to the suitability of each system as a means of communicating the relative sizes of earthquakes.**
   Magnitude is a quantitative measure of the energy of an earthquake and is measured using the Richter Scale whereas intensity often refers to the shaking effects of the earthquake and is estimated subjectively using the Modified Mercalli Scale.

5. **Countries like Australia only have relatively mild earthquakes. Why? Explain why nearby Papua New Guinea and New Zealand have major earthquakes.**
   Australia is situated well into its tectonic plate whereas New Zealand and Papua New Guinea are on the edges of plates (along a conservative fault and

subduction zone respectively). Australian earthquakes sometimes occur along old fault lines within the plate.

6. Briefly distinguish between:

   (a) **crust and mantle** – the upper layer of the Earth and the zone below respectively.
   (b) **seismograms and seismographs** – seismograms are the pattern of waves produced by the instrument called a seismograph.
   (c) **focus and epicentre** – the focus is the sources of the earthquake below the surface whereas the epicentre is the place immediately above it upon the surface.
   (d) **joints and faults** – joints are small to large cracks in rock whereas faults are joints along which movement has occurred.
   (e) **throw and heave of faults** – throw is the vertical displacement of a fault whereas heave is the horizontal displacement.

7. This question refers to the Travel-time graph for earthquakes given below:

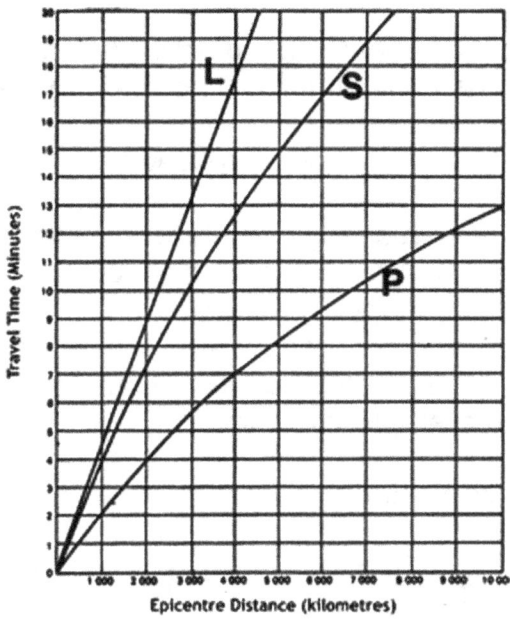

Three seismic stations received seismic wave patterns for the same earthquake:

STATION	P ARRIVAL TIME (HRS. MIN. SEC.)	S ARRIVAL TIME (HRS. MIN. SEC.)
A	11.22.00	11.25.00
B	11.23.00	11.27.00
C	11.26.00	11.30.30

(a) **How far is station A from the epicentre?** – with a P-S travel time difference of 3 minutes, the earthquake epicentre is about 2000 km away.

(b) **Give two reasons from the data why station A is probably closer to the epicentre?**
The P wave has arrived at the earliest time (11.22) and the P-S difference is the shortest (3 minutes)

8. **A table of recent earthquakes given previously in this book shows some very large earthquakes causing very few fatalities and some which are much smaller (remember the logarithmic scale of the Richter numbers) which have caused very large numbers of fatalities. Explain?**
The effects of the earthquakes at any habitable area depends on the ground structure and the nature of the human habitation. Greater fatalities are caused when the ground structure is relatively weak (especially poorly consolidated sediment often reclaimed) and the rigid structures built upon them collapse. In rural areas, the more flexible wooden buildings will withstand greater seismic vibration.

9. Use the internet to make a critical evaluation of the

   (a) earthquake hazard potential and
   (b) the tsunami hazard potential on your nearest coastline.

   Discuss this with your colleagues.

10. **Give examples of how research and observation from several scientific disciplines have led to an understanding of the nature of the Earth's interior and the Earth's shape.**
Seismic studies of P and S wave pathways have shown the interior to be layered. Magnetic surveys and studies of meteorites in astronomy have suggested a central nickel-iron core etc. See:

https://pubs.usgs.gov/gip/interior/

https://courses.lumenlearning.com/wmopen-geology/chapter/outcome-understanding-the-earths-interior/

https://study.com/academy/lesson/earthquakes-and-volcanoes-evidence-of-earths-inner-layers.html

https://www.bgs.ac.uk/discoveringGeology/hazards/earthquakes/structureOfEarth.html

https://www.appstate.edu/~marshallst/GLY1101/lectures/2-Earth_Composition_Structure.pdf

26.7 Reading List

The 10 Biggest Earthquakes in History. March 14, 2011, Australian Geographic.

NOAA, (2015): Interactive Map of Historical Tsunamis from NOAA's National Geophysical Data Center. http://maps.ngdc.noaa.gov/viewers/hazards/

Richter, C.F. (1958). *Elementary Seismology.* San Francisco: W.H. Freeman.

Skinner, B.J., Porter, S.C. & Park, J. (2004). *Dynamic Earth - An Introduction to Physical Geology.* 5th edition. 236pp: John Wiley & Sons, Inc.

Stein, Seth; Wysession, Michael (2009). An Introduction to Seismology, Earthquakes, and Earth Structure. Chichester: John Wiley & Sons. ISBN 978-1-4443-1131-0.

William H.K. Lee; Paul Jennings; Carl Kisslinger; Hiroo Kanamori (27 September 2002). *International Handbook of Earthquake & Engineering Seismology.* Academic Press. 283 pp. ISBN 978-0-08-048922-3

Chapter 27: Wind, Rain and Fire

27.1 Theme

There are many hazards associated with the atmosphere, hydrosphere and biosphere including hazardous weather patterns such as hurricanes, drought, floods and wildfires. These extremes of climate have a significant impact on communities, their economies and the varied natural environments of the country. An understanding of these extremes is most important in limiting destruction and making provision against such natural occurrences.

27.2 Rationale

With increased global warming, many of these extreme events have already begun to become reality. As this book is being written (February 2019), the north eastern part of Australia is undergoing the worst monsoon rain and flooding on record whilst in the south east of the country, a large part of the forested areas of the Great Dividing Range are experiencing widespread, uncontrollable wildfires with mass evacuations of local townships. Everyone needs to know about earth hazards; they are occurring with greater frequency and severity and they are largely unpredictable.

27.3 Notes on the Practical Work (see PRACTICAL MANUAL 2.)

Experiment 27.1: Runoff Coefficient and Flooding

This is another stream-table experiment which usually gives good results using basic equipment to model the real situation. The runoff coefficient (C) is a dimensionless coefficient relating the amount of runoff to the amount of precipitation received. In very general terms this means that the discharge of water across a surface (q) such as during flooding, depends upon the rainfall intensity (i) and the area it runs across (a) and the characteristics of the surface given by the Runoff Coefficient (C):

$$q \propto C i a$$

where q is the Peak Discharge;
C is the Runoff Coefficient;
i is the rainfall intensity; and
a is the area of the surface

The value of the ratio C also depends upon the characteristics of the surface such as its gradient, its soil type or covering material and any obstructions placed upon it.

Some experimentation is needed to get the initial angle and the amount of water right so that runoff can be measured without too much water flowing out of the tray. The measuring cylinder should be a large one but it might have to be replaced with a plastic container to hold the water coming from the tray. Precautions should be taken so that sand does not go down the sink. The apparatus could be set up outside or a plug could be put into the sinkhole. Carpet squares are ideal to simulate vegetation cover but they should not be of the deep pile variety. Any wooden blocks will do, provided that they are heavy enough to stand alone. On trials, blocks were of a denser wood and generally had a base of about 5 cm x 5 cm, but this is unimportant.

PART A: Basic runoff

QUESTIONS:

1. Why is it important to have a constant water flow and not allowing it to pool in the tank?
The water pool marks any structures seen in the sand.

2. Why should the carpet and the soil etc be wet before recording the water output in the measuring cylinder?
So as not to absorb any of the initial water flow.

3. What would be some improvements in the procedure to assist in better results?
A bigger or wider tank and a fine, broad shower head.

4. This experiment attempts to keep the rainfall as a constant. How could the effects of increased rainfall be measured?
Increase the flow of water which might give too much soil erosion so increase the number or size of the shower head.

CONCLUSIONS:

Runoff Coefficient concerns the ratio of the water flowing off a surface compared to the rainfall falling upon it. Comment on the factors which will affect this ratio assuming that the rainfall is constant.
These include: the type, amount, intensity and the duration of the precipitation (rain, snow, sleet, etc.) and the direction of the storm which produced it.

RESEARCH: (Optional)

Use the internet to find out more about Runoff Coefficient, how it is measured and how it is used to reduce floods.

See:

https://www.hydrol-earth-syst-sci-discuss.net/9/4919/2012/hessd-9-4919-2012.pdf
(Click on the boxes at right for more information)

http://arr.ga.gov.au/__data/assets/pdf_file/0017/40553/ARR_Project_13_Stage3_report_DRAFT.pdf

https://ewater.org.au/archive/crcch/archive/pubs/pdfs/industry199805.pdf

27.4 Other Activities

- Photograph the clouds each day for a week and compare them to a set of cloud types to identify them. See the textbook and:

    http://australiasevereweather.com/techniques/moreadv/class.htm

    https://cloudatlas.wmo.int/cloud-identification-guide.html

- Make a model 'tornado'. See:

    http://www.tornadoproject.com/cellar/workshop.htm

    http://onetimethrough.com/make-hurricane-jar/

- Demonstrate the Coriolis Effect. See:

    https://www.ducksters.com/science/experiment_coriolis_effect.php

    https://www.carolina.com/teacher-resources/Interactive/modeling-the-coriolis-effect/tr10643.tr

    http://www.cosee.net/cosee-west/Feb2012/Coriolis_El%20nino.pdf

- Examine a local flood potential map and notice the possibility of future flooding in the area. Do this for several world locations e.g. several major coastal cities such as Brisbane, Sydney, Melbourne, San Francisco, Dhaka, Djakarta, London, New York, Mumbai, Manilla, Nagoya and many Pacific Islands such as Kiribati.

    http://www.floodmap.net/

    http://globalfloodmap.org/

    http://floodinformation.brisbane.qld.gov.au/fio/
    (Brisbane)

- Demonstrate the effects of oxygen and carbon dioxide gases separately on fire by making test-tubes of these gases and inserting a lighted taper.

- Demonstrate lightning with a van der Graaf or another electrostatic generator. Hold a small piece of hay or similar to see how lightning can set fire to timber.

27.5 Answers to Multichoice Questions. See:

Q1.B   Q2.C   Q3.A   Q4.D   Q5.B   Q6.C   Q7.B   Q8.A   Q9.D   Q10.C

27.6 Some Suggestions for the Review Questions

1. **Compare and contrast the El Niño and La Niña effects showing how they are caused and what are their outcomes. Why are they important to the climate of Pacific countries?**
   They concern the movement of warm pools of water and moist air across the central Pacific due to the influence of the prevailing Trade Winds. During an El Niño, the pool of warm water is in the eastern Pacific bringing rains to the Americas but drought to the western Pacific. See:

   https://oceanservice.noaa.gov/facts/ninonina.html

2. **What are some natural indications that a storm may be developing or approaching? Research the truth about some weather lore to indicate approaching storms such as flocks of birds fleeing the area etc.**
   Often depends upon local conditions. A long front of cloud, rapidly moving towards the observer may mean a storm front approaching. A sudden drop in air pressure in a barometer is also a good sign of a storm. Clouds building into thunderheads are also another warning. Some people also look for warning sign in animal behaviour such as ants climbing walls, birds flying away from the area. Often high, wispy alto cirrus clouds may herald unstable weather and rain within a few days. See:

   https://blogs.unimelb.edu.au/sciencecommunication/2012/10/25/weather-lore/

   https://www.spinnakersailing.com/weather-lore/

3. **What is meant by each of the following terms:**

   (a) **Warm front** – when a mass of warm, moist air moves in replacing cold air usually bringing stratiform clouds and rain.
   (b) **Squall lines** – form in unstable air often forward of a cold front bringing fast, turbulent storms.

- (c) **Isobars** – lines joining places of equal air pressure on weather maps.
- (d) **Saffir-Simpson scale** – is a hurricane wind scale using a 1 to 5 rating based on a hurricane's sustained wind speed. This scale estimates potential property damage.
- (e) **Flood stage** - is the level at which a body of water's surface has risen to a sufficient level to cause sufficient inundation of areas that are not normally covered by water.

4. **What is the Coriolis Effect? How does it affect the winds of the world?**
   This is the apparent deflection of a body in motion such as the wind with respect to the Earth, as seen by an observer on the Earth, caused by the rotation of the Earth and appearing as a deflection to the right in the Northern Hemisphere and a deflection to the left in the Southern Hemisphere.

5. **What are the main types of flooding which can occur? How does flooding generally affect either (a) a rural community or (b) an urban community?**
   Flooding can be due to rivers, storm surges or sheet runoff across flat land. In rural communities much damage is done to crops, buildings and livestock which may be trapped in sudden flooding. In severe circumstances, topsoil and infra-structure may also be washed away. In urban areas there is mass inundation of streets and buildings, communications and power usually fail and transportation systems come to a standstill. See:

   http://www.bom.gov.au/australia/flood/EMA_Floods_warning_preparedness_safety.pdf

   https://www.usgs.gov/faqs/what-are-two-types-floods?qt-news_science_products=0#qt-news_science_products

6. **What is meant by each of the following terms? A diagram may help in the explanation:**

   - (a) **Lag time** – is the delay between when rainfall occurs and when the discharge of the river actually increases and is measured as the time between the middle of the rainstorm and the time when discharge reaches its peak.
   - (b) **Flood peak** – the highest level of water.
   - (c) **Catchment area** – area of land drained by a creek or river system.
   - (d) **Levee bank** – raised river bank built up along with the bed above the surrounding flood plain by sediment deposition.

(e) **Overbank flow** – is water flow that is greater than the river channel can handle and so the water flow goes over the river's banks and onto the flood plain.

7. **What are canopy fires? What factors determine the speed and spread of such fires? Why are they considered to be very dangerous?**
These are fires that burn the whole forest canopy as a single entity, which include crown fires in the tops of individual trees. They are very dangerous because the wildfire can spread very quickly over the top of a forest sending fireballs and sparks many kilometres. They also can jump from one area to another making firefighting and containment difficult.

8. **Use the Internet to research some of the most recent and major community disasters associated with:**
    (a) Extreme storm events
    (b) Flooding
    (c) Wildfires

Requires individual searches for each of the categories listed above. Perhaps selected tasks could be delegated with research followed by class discussion.

9. **What are the major hazards associated with wildfires? How can they be prevented or overcome?**
Apart from actual incineration and heat radiation, there are also hazards associated with smoke, carbon monoxide and dioxide poisoning and hazards due to falling buildings, exploding fuels and gas and general panic. See:

http://myfirewatch.landgate.wa.gov.au/
(a good interactive map of Australia for bushfires)

https://pubs.usgs.gov/fs/2006/3015/2006-3015.pdf

10. **What precautions should one take in preparation for:**

    (a) an impending cyclone/hurricane season
    (b) local extreme storms during summer
    (c) extensive rain with the possibility of flooding
    (d) potential wildfires during a very hot, dry summer.

Each disaster has its own unique precautions and should be researched separately. A good site about preparation for natural disasters is found at:

https://www2.health.vic.gov.au/emergencies/emergency-type/natural-disasters

https://www.qfes.qld.gov.au/community-safety/documents/GetReadyGuide-E.pdf

27.7 Reading List

Abbott, P. L. (2009). *Natural Disasters*. 7th edition. Dubuque, IA, McGraw-Hill. ISBN: 9780078022982.

Baumgarner, J.B. (2008). *Emergency Management: A Reference Handbook* Santa Barbara: ABC-Clio publications. ISBN: 978-1-59884-110-7.

Gunn, A. (Edit) (2007). *Encyclopedia of Natural Disasters-Environmental Catastrophes and Human Tragedies*. Santa Barbara: ABC-Clio publications. ISBN: 978-0-313-34002-4.

Haddow, G., Bullock, J. A., Haddow, K. (Edits) (2009). *Global Warming, Natural Hazards, and Emergency Management*. Boca Raton, Florida: CRC Press: ISBN: 9781420081824.

Paron, P. (2014). *Hydro-Meteorological Hazards, Risks, and Disasters*.1st Edition. Amsterdam. Elsevier.ISBN: 9780123948465.

Paton, D. (Edit.) (2014) *Wildfire Hazards, Risks, and Disasters*. 1st Edition. Amsterdam: Elsevier.ISBN: 9780124096011.

Shi, P. (2019). *Disaster Risk Science*. Singapore, Springer. ISBN: 978-981-13-1851-1

Smith, K. (2009). *Environmental hazards: assessing risk and reducing disaster*. 5th ed. New York, NY: Routledge. ISBN: 9780415428637.
https://arjzaidi.files.wordpress.com/2016/09/9781136647154_sample_8306701.pdf

Chapter 28: A Changing Climate

28.1 Theme

The world's climate changes throughout time by natural processes and these ancient climates are studied by the science of palaeoclimatology. The current debate is about anthropogenic climate change; that rapid and extreme set of processes fuelled by the activities of humankind which are changing aspects of the climate to threatening levels. There is considerable evidence of climate change from a variety of methods and scientific disciplines all including that the Earth's atmosphere and hydrosphere are warming with often harmful consequences.

28.2 Rationale

Climate change with global warming has been occurring since the beginning of the Industrial Revolution but more dramatically over the last fifty years. The consequence of climate change, including increased ocean and land temperatures, more severe storms and degrading of many habits, are already occurring. It will take a massive change in attitude, life-style and improved industrial and political management if this trend is to be reversed. That is the point of this book!

28.3 Notes on the Practical Work (see PRACTICAL MANUAL 2.)

<u>Experiment 28.1</u>: **Correlation of Carbon Dioxide Levels and Global Temperature Change**

This is a good introduction to those students who have not had any experience with statistics and their use in finding trends. The basic equation for Pearson Correlation Coefficient is given in the introduction but this should not be necessary as students will be using an online calculator for this. In detailed research this is what happens; data is gathered by the researcher and then fed or coded into a computer program which then does the appropriate calculations and gives a numerical result which confirms or not the hypothesis being tested. This is a simple test of correlation between carbon dioxide levels and global temperatures. Worldwide analysis of such parameters has shown a strong positive match as should this experiment.

The correlation calculator is found at:

> https://www.socscistatistics.com/tests/pearson/Default2.aspx

It is simple to use; student simply type in the separate columns given as data and then hit the CALCULATE R button. Some other good statistics are given as well as the correlation value between 0 and 1 with 1 being an exact match.

QUESTIONS:

1. Why are temperature anomalies used instead of maximum or minimum temperatures?
Maximum and minimum values can be misleading and may vary slightly due to local conditions. Temperature anomalies are used against known averages.

2. Why is the data given as concentration of carbon dioxide in parts per million rather than as an emission value in tonnes?
So that there is no confusion between what happens to the emissions and other interactions.

3. Why are CO_2 concentrations measured on top of a volcano in Hawai'i? What problems are associated with measuring CO2 concentrations?
There would be other, more local concentrations which would not be representative of the global atmosphere e.g. measuring the values in the city would give additional carbon dioxide concentrations for industry and the population.

4. Why is statistical analysis such as the Pearson Correlation Coefficient used?
It is a good way of handling large amounts of data and also to be able to make a comparison between one set of data and another.

5. Are there any problems or errors associated with this type of calculation (some Internet research needed)?
These tests assume random sampling and that the statistics are normally distributed and that the test assumes that the correlation coefficient is calculated on interval or ratio data. See the following for more details:

https://ncss-wpengine.netdna-ssl.com/wp-content/themes/ncss/pdf/Procedures/PASS/Pearsons_Correlation_Tests.pdf

CONCLUSIONS:

1. What was the correlation coefficient between atmospheric carbon dioxide concentrations and global temperature?
The value of R is: 0.9887, a very good match.

2. What does this show about the relationship between these two variables over time? Discuss.
That there is a direct linear relationship between temperature anomalies and carbon dioxide concentrations.

RESEARCH: (Optional)

Use the internet to find out more global warming and carbon dioxide gas concentrations and emissions, especially the emissions of different countries. See:

https://ourworldindata.org/co2-and-other-greenhouse-gas-emissions
(which gives interactive maps showing data for most countries. Compare the data for Australia, United States, China and some of the less-industrialised nations.)

http://berkeleyearth.lbl.gov/city-list/
(also gives the global warming data for major cities.)

Also bookmark the following site which gives a daily measure of global carbon dioxide levels:

https://www.co2.earth/daily-co2

28.4 Other Activities

1. Make a carbon dioxide generator using a large test-tube, cork and double right-angled delivery tube. Use acid and marble to generate the gas and run it into a second test-tube of water with Universal Indicator. Notice the development of acidity as carbon dioxide is passed into the water.

2. Set up two identical large round flasks with thermometers through their stoppers. Fill one with carbon dioxide and leave air in the other. Put them both together in front of an infra-red lamp and see if there is any difference between the temperatures in both flasks.

3. Build a small Coolgardie Safe by using a small box frame of timber and hessian bag with a tray of water on top and dripping strips hanging down each side. An old wooden supply box would be useful. Put it in a draughty area and measure the temperature each day.

4. Set up a simple hydroponics system in a sunny window. Use sterilized sand, perlite, crushed and washed gravel or any other inert substance as the holding matrix. Use a total plant food purchased from a supermarket and water by hand every two days or have a drip system from a tank needing only regular water addition. Grow the plants in old troughs or pots standing in trays. Plant fast-growing wheat or chives or spring onions from seeds soaked in warm water over night.

5. Discuss potential lifestyles of the latter 21st century assuming reliance on renewable energies, building resources and styles, food production, climate control (no air conditioning), mass rather than individual transportation.

28.5 Answers to Multichoice Questions

Q1.C Q2.A Q3.D Q4.A Q5.C Q6.A Q7.B Q8.C Q9.D Q10.B

28.6 Some Suggestions for the Review Questions

1. **There is still considerable opposition to the notion of climate change and the need to change aspects of modern life. What are the main points proposed by antagonists to climate change and why do you think they put forward?**
   The few arguments left deny that there is any increase in temperature levels and that carbon dioxide concentrations are not a reliable influence. Mostly the arguments seem to be wanting no change and, especially wanting cheap electricity through coal-fired power stations. One must question such arguments as there may be some bias with those working in energy and mining industries.

2. **Describe what is meant by the following terms:**

   (a) tipping point - A climate tipping point is a point when a global climate changes from one given stable state to another stable state.
   (b) nuclear winter – an old concept based on the idea that a nuclear war would produce colder climates because of the dust covering the atmosphere. It has been applied to any system which will prevent heat from the Sun reaching the Earth's surface.
   (c) oxygen-18 isotope – is one of the isotopes of oxygen. In ice cores, mainly Arctic and Antarctic, the ratio of ^{18}O to ^{16}O can be used to determine the temperature of precipitation through time.
   (d) albedo – the reflectivity of the Earth's surface and clouds.
   (e) permafrost – permanently frozen soils in high latitudes.
   (f) algorithm – a computer mathematical program.

3. **Why have scientists concerned with climate change studied the great ice ages of the past? What research methods have they used?**
   They were attempting to find whether or not the current changes in climate e.g. atmospheric temperature increase was seen in the past as a natural part of interglacial warming. They used similar methods to today including ice cores, sea sediment cores, microfossils, glacier positions etc.

4. **Suggest why plate tectonics could also be used to counter arguments for climate change?**
   It could be argued that recent climate change is not due to Human-kind's activities but natural phenomena due to long-term processes of plate tectonics such as opening of oceans, volcanic eruptions and movement of plates into different climatic zones.

5. **Evidence for climate change has come from a great many scientific disciplines. List these sciences and give a few words explaining the type of evidence that they have acquired in measuring climate change.**
   These would include:

Climatology - oxygen ratios in ice cores
Palaeontology - micro- and other fossils
Geology - plate tectonics and volcanoes
Astronomy - Earth's tilt and movement, relationship with Sun's radiation
Oceanography - measurements of ocean temperatures
Meteorology - monitoring weather patterns and atmospheric gas concentrations etc.

6. **The scientific community is concerned about a two degree increase in climate. This does not seem much so why the concern?**
This would be an averaged value so some parts of the world will have a much hotter climate than just an extra 2 degrees. Also, even if taken as an average this is a lot of heat increase. Remember that temperature is the rise in heat of a mass - and the Earth's surface contains a lot of mass! They have also noticed the effects in the distant past when global temperatures went up or down only a few degrees and noticed great changes.

7. **List the main findings which the global scientific community has found from their research into climate change.**
This will be a summary of the graphs found in this chapter e.g. carbon dioxide levels have risen to past 400 ppm, global air temperatures have risen, global sea temperatures have risen, ice caps and glaciers have been reduced etc.

8. **Describe the main factors involved in the Milankovitch Cycle and evaluate their importance in the increase in global temperature. What other factors can cause natural climate change?**
These factors include: Earth's orbital shape; its axial tilt; its axial precession; its orbital precession; and its orbital inclination to the Sun. Other factors would include the intensity of the Sun and any change in the distance from it. There are other factors include those on Earth such as volcanic eruptions, plate movements and albedo.

9. **Consider the immediate local environment. What effects will an increase in temperature with hotter summers and milder winters with an increase in storms have on the local ecology?**
This would concern the immediate environment of the location of the readers of this book. One must consider the changes in average temperatures through the seasons if there is a temperature increase, the change in precipitation, especially if rainfall is reduced or there are more frequent and violent storms. Consider the conditions of housing, obtaining food and water, local movement and transportation and general lifestyle.

10. **Discuss what a future lifestyle in an urban region might be like in the future:**

    (a) **with hotter summers, milder winters and a generally drier climate** - hotter summers would immediately suggest a greater reliance on air conditioners but this would mean greater energy use and in countries using coal-fired power stations this becomes a circular argument. Cooling using natural processes such as open-plan buildings with shady verandas and window shades, use of more running water features within buildings, changes to daylight work/play options etc. A drier climate will put more stress on current water supplies so further improvements in water storage such as additional dams (also to be used with pumped hydroelectricity), water storage internally in all buildings collecting from the roof, solar desalination plants, limitations on water use etc.

(b) **with the removal of fossil fuels and nuclear power** – a change over to the use of renewable power such as individual and industrial solar electricity, wind power, hydroelectricity etc. This will also require changes in the use of transport and generally a less-wasteful lifestyle.

(c) **higher sea levels and more extreme storms** – flooding and inundation of coastal and low-lying river flood plains will mean a shift in population to higher ground. This could also have vast socio-political implications if this shift is across international borders.

(d) **new technologies in transportation, personal climate control and housing** – see above but also a change from individual transportation to mass public transportation using renewable energy systems e.g. high-speed rail links between dormitory suburbs or cities and decentralized work places.

28.7 Reading List

Flannery, T. (2015). *Atmosphere of Hope- Searching for Solutions to the Climate Crisis*. New York: Atlantic Monthly Press. ISBN: 9780802124067.

Hawken, P. (Edit) (2017). *Drawdown: The Most Comprehensive Plan Ever Proposed to Reverse Global Warming*. New York: Penguin Books. ISBN: 9780143130444.

Kolbert, E. (2015). *The Sixth Extinction – An Unnatural History*. New York: Henry Holt & Co. ISBN: 978-0-8050-9299-8.

McKibben, B. (2011). *Earth: Making a Life on a Tough New Planet*. Black New York, Henry Holy & Co. ISBN: 978-0-8050-9056-7.

Romm, J. (2015). *Climate Change: What Everyone Needs to Know*. New York: Penguin Putnam Press. ISBN: 10 0143130447.

Vince, G. (2014). *Adventures in the Anthropocene: A Journey to the Heart of the Planet We Made.* Minneapolis, Minnesota, Milkweed Editions. ISBN: 1571313575.

APPENDIX A: Risk Assessment of Practical Work and Excursions

It has become mandatory that a risk assessment is made before carrying out any laboratory practical activity or excursion. These must consider:

1. the nature of the activity;
2. equipment, chemicals, living things and other items to be used;
3. the nature of the environment, especially if venturing outside for an excursion, even in the local area; and
4. the student.

Once all of the factors and options have been considered, a Risk Assessment Form should be completed and kept in a central records section (often kept by a Laboratory Manager). It is also a good idea for a copy to be kept (electronically and print) by the teacher along with the set of notes for the experiments and activities (perhaps attached to the teacher's copy of this book). Once completed, it should be revised and modified as required each time the activity is performed.

This need not be an onerous task after it has been done for the first time. Thereafter it simply requires a quick review of the previous form and modifications if any of the variables (e.g. different student type, different location, new equipment, different teacher etc.) have changed.

Safety is paramount for any teaching activity and the students must be made aware of any potential hazard in the activity and take adequate precautions. Some suggestions for laboratory safety and safety during field excursions have been given at the beginning of this book. It is always useful to have these suggestions displayed prominently in the laboratory. In the textbook ADVENTURES IN EARTH and ENVIRONMENTAL SCIENCE, each chapter contains a PRACTICAL TIPS at its conclusion. These have been based on over forty years of safe laboratory and field practices as a field researcher and many years travelling into some of the most hostile and remote parts of the worlds, such as the Antarctic Peninsula, Amazon Basin, North African deserts and alpine regions in New Zealand, Europe and the Andes. This section also outlines, as do various, parts of the body of the textbook, how science is one of the most interesting endeavours of Humankind. Institution laboratory managers should have on file a detailed indexed catalogue of the risk or hazard potential of all chemicals, electrical equipment, biological specimens (including live organisms) and other items in use within the
laboratories.

A simple Risk Assessment Form is given on the next page:

Earth and Environmental Science Risk Assessment Form		
DATE: CLASS: LOCATION:		
ACTIVITY: TYPE (Pract./Excursion/visit):		
RISK LEVEL: Students: Teacher: Other(who/what): (High/Medium/Low)		
OUTLINE of ACTIONS	**RISK**	**PRECAUTION/ACTIONS**
DISPOSAL of WASTES:		
OTHER COMMENTS:		

A more detailed Risk Assessment Form can be found at:

https://www.aisnsw.edu.au/workplace-health-and-safety/Documents/Appendix_A_Science_and_Technology_Risk_Assessment_Template_Rev.1.docx

Some useful references for risk assessment can be found at:

https://education.qld.gov.au/sitesearch/Pages/results.aspx#k=risk%20assessment

https://education.qld.gov.au/initiatives-and-strategies/health-and-wellbeing/workplaces/safety/managing/risk-management

https://www.riskassess.com.au/info/routine_safety_procedures

https://smah.uow.edu.au/content/groups/public/@web/@sci/@chem/documents/doc/uow016874.pdf

https://www.riskassess.com.au/docs/RABrochureAU.pdf

http://www.nswtitration.com/files/school_risk_assess.pdf

https://assist.asta.edu.au/

https://assist.asta.edu.au/search?expert=&field_curriculum_year=All&field_publication_date=All&keywords=risk+assessment&field_tax_australian_curriculum_parent_parent_parent_tid=&laboratory_technicians=All&area=&field_voting_user_rating=&field_voting_average_rating_1=&field_voting_user_rating_1=&rating_point=All&year_level=&sort_by=created

https://education.nsw.gov.au/teaching-and-learning/curriculum/key-learning-areas/science/safety

https://www.education.vic.gov.au/school/principals/spag/governance/Pages/riskinplanning.aspx

http://ascip.org/wp-content/uploads/2016/06/ASCIP-Risk-Management-Primer-for-School-Districts-SIXTH-DRAFT-2016-06-20.pdf

https://www.rospa.com/rospaweb/docs/advice-services/school-college-safety/managing-safety-schools-colleges.pdf

https://www.leeds.ac.uk/secretariat/documents/risk_management_guidance.pdf

https://www.teachers.org.uk/files/safety-in-practical-lessons_0.doc

APPENDIX B: Excursion Permission Note

Some institutions often require an Excursion Permission Note from students who are under legal age who are leaving the institution for a field, industrial/scientific visit or other outside activity. These DO NOT absolve the teacher and guides of any moral and legal responsibility but merely provides parents/guardians with the necessary information as to what their child will be doing and where. It also can provide a list of safety and comfort items which will be needed from home for the activity in the hope that such reasonable items will be supplied.

Some examples of field or other outside activities for this program include:

- Rock quarries
- Mining areas (special permission and rules from the company)
- Freshwater stream ecology
- Marine rock platform ecology
- Rainforest/dry forest/grasslands ecology
- Mining/mineral museums
- Environmental stations (special rules and guides apply)
- Museums of natural history
- Scientific research organisations (special rules and guides apply)
- National parks (special local rules and guides necessary)
- Government agencies (special rules and guides apply)

A typical activity permission note is given on the next page and may be copied for the students to take home:

< insert institution letter head>

EARTH and ENVIRONMENTAL SCIENCE EXCURSION

DATE: <insert date and times>   CLASS: < insert class/group>

DESTINATION: <insert location/name of place or organisation etc.>

This excursion is an integral part of the semester's programme. All students will be required to complete an assignment, associated with the excursion, which will contribute towards the assessment of the subject.

ARRANGEMENTS: < insert details of transportation, time of departure and return etc.>

The bus will leave at   < time>   and returns to the school at   < time> approx.
      (Bus company can be contacted at: <name and contact number>   )

Students are to meet at: < meeting place and time>
but will not to enter the bus until directed to do so. The rules of the School apply at **all times**

Staff going on the excursion will be: < name(s) of staff and contact numbers >

SPECIAL REQUIREMENTS: < special requirements such as dress, personal items such as cameras, mobile phones, writing material, safety precautions etc. Also, any special rules of behaviour and group actions if lost etc.>

Yours faithfully,
 < name of person in authority>
Head of Earth and Environmental Science

CONSENT FORM: Please return to supervising teacher by   < insert time/date>
I have read the above information and agree for...... <insert student name>
........to go on the excursion.
Special information concerning my daughter's/son's welfare that the supervising staff should know is as follows:

<insert special requirements such as allergies, dietary requirements, medical conditions etc.>

Phone contact of parent/guardian

Home:...........................Work...............................

Signature:...(Parent/Guardian)

Date ……………………………….

Errata from the First Edition

<u>Book 1:</u>

1. Page 357, Chapter 11: Auto numbering error on Question 2 so there are only nine questions. An additional question could be:

   10. This oxygen release also allow for the formation of:

      A. Free nitrogen from nitrate minerals in the sea

      B. Free carbon dioxide gas from carbonates in the sea

      C. Free carbon dioxide gas from methane in the air

      D. Free nitrogen from organic compounds in the sea

   The answer is C.

2. Page 358, Chapter 11: Question 10, derived not drived.

3. Page 428, Chapter 13: Auto numbering in the wrong place so Question 9 is Q.8 and Question 10 is Q. 9. There is no question 10. A new Question 10 could be:

   10. The inventor of the radio telescope was:

      A. Carl Sagan
      B. Karl Jansky
      C. Jerry Ehman
      D. Frank Drake

   The answer is B.

4. Page 457, Chapter 13: Question 6 d of the Review Questions should read Advection not aduction.

5. Page 494, Chapter 14: Multichoice Question 2d should read Advection not aduction.

6. Page 496, Chapter 14: Review Questions 4: replace their with there.

7. Page 496, Chapter 14: Review Questions 5c: Replace Blackmann's with Blackman.

8. In Practical Manual 1. Experiment 14.4: The numbering system has not been consistent.

Book 2:

1. Page 29, Chapter 15: Multichoice Questions: There are two Question 1s. No action required.

2. Page 129, Chapter 19: Multichoice Questions: No Question 5. In the Second Edition it would be:

    5. The open space left behind after a section of mine has been excavated is called a:

    A. Bord
    B. Stope
    C. Cave
    D. Level

    The answer is B

3. Page 298, MULTICHOICE QUESTIONS 10, Missing diagram. Should be:

This is Wilpena Pound, an eroded dome in South Australia so the answer is C.

Practical Manual 1.

Experiment 2.2, P. 22. No numbers on PROCEDURES

Experiment 3.2, P. 29. PROCEDURES numbering continues at 5 not 4.

Experiment 10.3, p. 95. Add numbering missed to PROCEDURE.

Experiment 13.2, p. 118. PART C: PROCEDURE not PROCEEDURE

Experiment 14.2, p. 125 numbering should continue as 5 not 6.

Practical Manual 2.

1. Experiment 17.1, Page 16. No Conclusion. Students should find this obvious and write their own from the Questions.

2. Experiment 17.2, p.17 RESULTS has been duplicated.

3. Experiment 18.1, p. 25 QUESTIONS should be renumbered from 1 not 7.

4. Experiment 18.2, p 28 PROCEDURE has two part 2. notations. Change the second one to 3.

5. Experiment 21.7, p. 74 PROCEDURE has poor numbering not starting at 1.

6. Experiment 22.1, p. 78 QUESTIONS 4 and 5: the websites do not open so use

    https://www.pewtrusts.org/-/media/assets/2017/10/story_of_atlantic_bluefin_tuna_science.pdf

7. Experiment 22.1, Part B, p. 78 QUESTION 6 "What factors caused…"

8. Experiment 22.2, p. 81 QUESTION 3 should read " in Question 2c" not 10c.

9. Experiment 22.5, p. 90 QUESTION 5c should read in brackets "(some additional **research** may be needed)"

10. Experiment 23.2, p 100 top of the page, the number should be 7 not 4. And in Question 3 it should read " …parallel connect**ion**. On page 101, numbers should have been continued from 4 onwards.

11. Experiment 23.3, p. 104 RESEARCH should read "…is current**ly** used.."

12. Experiment 23.4, p. 106, QUESTIONS 7 should read '..how was **the** condensate tested .."

13. Experiment 27.1, p. 132 PROCEDURES. Delete Part A – there is only one procedure. Also p.134, QUESTIONS numbering has continued from previous section. Start at 1 etc. on that page.

14. Appendix A: Risk Assessment, third last paragraph, fifth sentence should read '…. a field researcher **with** many years travelling…' not 'well as'.

Books by the Author

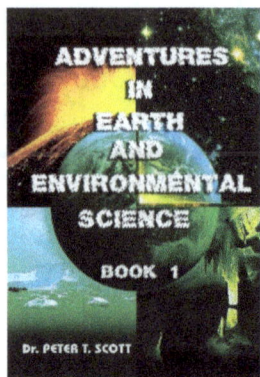 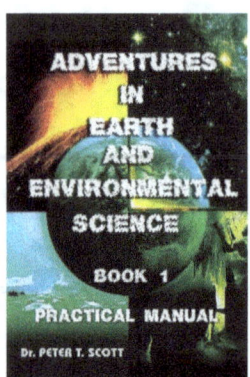

ADVENTURES IN EARTH AND ENVIRONMENTAL SCIENCE - BOOK 1. This is the first book of two which looks at the Earth, its matter, energy relationships and its life and how all interact together as a whole. The Earth is seen as a closed system contain the Earth's materials and living things but allowing a necessary flow of energy into and out of the planet. The atmosphere, hydrosphere, geosphere and biosphere of Earth are all examined in detail and lavishly illustrated with over five hundred photographs and diagrams. There are also several links to videos made by the author during his own adventures in studying the Earth. Each chapter is concluded with a Summary, Practical Tips, ten Multichoice Questions and ten longer Review and Discussion Questions. There is also an accompanying PRACTICAL MANUAL with a large number of experiments and data analysis activities which can be performed by students using basic available equipment in support of the textbook. This manual also teaches students how to investigate and write research reports for submission.

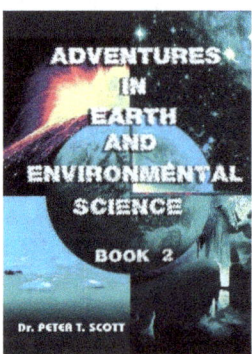 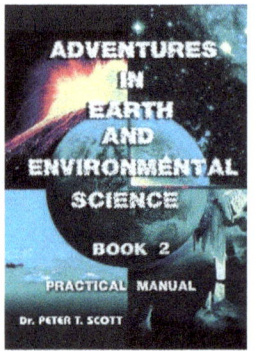

ADVENTURES IN EARTH AND ENVIRONMENTAL SCIENCE - BOOK 2. This is the second book with this title and, after a short revision chapter, looks at how Humankind lives on planet Earth. Renewable and non-renewable resources are described and how their use has impacted on the world's ecosystems, on land, in the sea and in the atmosphere. This is also discussed with an emphasis on the problems of future energy needs, global warming and social consequences of these events. As well as problems caused by Humankind, the natural hazards of the Earth have also been described with the view that many or the world's populations live in regions that can be very dangerous at times. The contents of this book are also supported with many photos, illustrations and videos. Each chapter is concluded with a Summary, Practical Tips, ten Multichoice Questions and ten longer Review and Discussion Questions. Book 2 also has an accompanying PRACTICAL MANUAL with a large number of experiments and data analysis activities which can be performed by students using basic available equipment in support of the textbook.

ADVENTURES IN EARTH AND ENVIRONMENTAL SCIENCE
This is the composite book containing all of the content of Books 1 and Books 2. It has been written as a utilitarian reference book for the classroom, library or home study.

ADVENTURES IN EARTH AND ENVIRONMENTAL SCIENCE-TEACHERS' GUIDE
This has been designed to assist teachers with the use of these books and the teaching of this subject. It gives advice on lesson preparation, the teaching of the practical work and answers to the questions contained in the books.

Adventures in Earth and Environmental Sciences Books 1, 2, the composite book and the Teachers Guide are all available in electronic (Kindle) format which can be viewed using any electronic device having the free Kindle App. They are available with Readcloud and soon with other servers in the school systems. They are also available in PRINT editions from Felix Publishing at (info@felixpublishing.com)

Other books in the **ADVENTURES** series follow a more traditional Earth Science content in the sciences of Geology, Oceanography, Meteorology and Astronomy. They are also available in electronic format from Kindle as well as in print form direct from Felix Publishing (info@felixpublishing.com) and include:

ADVENTURES IN EARTH SCIENCE ADVENTURES in EARTH SCIENCE is an in-depth textbook as well as a series of adventures across seven continents and beyond in the sciences of astronomy, geology, meteorology and oceanography. It has been written with over forty years of experience in studying, researching and teaching earth science. Whilst it has been designed for senior high school and junior university or college, it is written in an easy style and well-illustrated so that anyone with an interest in this topic would find it an interesting and valuable resource.

This reference book has also been reprinted in the smaller A5 size as separate topic editions for easier reading by anyone wishing to have the most up to date information on these topics. The smaller books are:

EXPLORATION SCIENCE
Field Geology & Mapping

RICHES from the EARTH
Minerals & Energy

CHANGING the SURFACE
Erosion & Landscapes

ROCKS - BUILDING the EARTH

  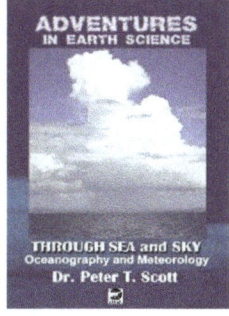 

A DANGEROUS PLANET
Volcanoes & Earthquakes

FOSSILS - LIFE in the ROCKS. Fossils and past environments

THROUGH SEA and SKY
Oceanography & Meteorology

BEYOND PLANET EARTH
An Introduction to Astronomy

For more information about all of these books and world-wide distribution contact the publisher direct at info@felixpublishing.com

www.ingramcontent.com/pod-product-compliance
Lightning Source LLC
Chambersburg PA
CBHW050713090526
44587CB00019B/3362